IN THE STORMS OF TRANSFORMATION

Two Shipyards between Socialism and the EU

In the 1990s, states in what would become the eastern edge of the European Union transformed their political systems and economies, leaving state socialism behind for liberal democracies and free markets. In the ensuing decades, two shipyards that were once the pride of their cities – in Gdynia, Poland, and Pula, Croatia – went bankrupt, unable to withstand global competition.

Through an interdisciplinary study of these two shipyards, *In the Storms of Transformation* brings together a team of researchers to re-evaluate the shift from state socialism to market capitalism and offer a new periodization. With perspectives from social anthropology, sociology, and business history, the book argues that this transformation began with the oil crisis of the early 1970s and ended with EU accession – in 2004 in Poland and in 2013 in Croatia – highlighting the EU competition laws and global competition that pushed the shipyards into bankruptcy and diminishing the role of the revolutions of 1989.

In the Storms of Transformation bridges local labour history with global market forces, going beyond prevalent narratives of loss and nostalgia or successful neoliberal change to offer a novel and nuanced reading of post-communist transformation and its contradictions.

(German and European Studies)

ULF BRUNNBAUER is the academic director of the Leibniz Institute for East and Southeast European Studies and holds the Chair of Southeast and East European History at the University of Regensburg.

PHILIPP THER is a professor of Central European history and founder of the Research Center for the History of Transformations (RECET) at the University of Vienna.

PIOTR FILIPKOWSKI is an assistant professor at the Institute of Philosophy and Sociology at the Polish Academy of Sciences.

ANDREW HODGES is a book editor and literary translator at The Narrative Craft.

STEFANO PETRUNGARO is an associate professor in the Department of Linguistics and Comparative Cultural Studies at Ca' Foscari University of Venice.

PETER WEGENSCHIMMEL is head of archives at the University of Kassel.

GERMAN AND EUROPEAN STUDIES

General Editor: James Retallack

In the Storms of Transformation

Two Shipyards between Socialism and the EU

THE SHIPYARD COLLECTIVE

Ulf Brunnbauer
Philipp Ther
Piotr Filipkowski
Andrew Hodges
Stefano Petrungaro
Peter Wegenschimmel

Translated by Carla Welch

UNIVERSITY OF TORONTO PRESS
Toronto Buffalo London

© University of Toronto Press 2025
In den Stürmen der Transformation: Zwei Werften zwischen Sozialismus und EU
© Suhrkamp Verlag Berlin 2022
All rights reserved by and controlled through Suhrkamp Verlag Berlin.

Toronto Buffalo London
utorontopress.com

ISBN 978-1-4875-5032-5 (cloth)   ISBN 978-1-4875-5037-0 (EPUB)
ISBN 978-1-4875-5034-9 (paper)   ISBN 978-1-4875-5035-6 (PDF)

German and European Studies

___

Library and Archives Canada Cataloguing in Publication

Title: In the storms of transformation : two shipyards between socialism and the EU / Ulf Brunnbauer, Philipp Ther, Piotr Filipkowski, Andrew Hodges, Stefano Petrungaro, and Peter Wegenschimmel; translated by Carla Welch.
Other titles: In den Stürmen der Transformation. English
Names: Brunnbauer, Ulf, 1970–, author.
Series: German and European studies; 56.
Description: Series statement: German and European studies; 56 | Translation of: In den Stürmen der Transformation : zwei Werften zwischen Sozialismus und EU. | Includes bibliographical references and index. | In English, translated from the German.
Identifiers: Canadiana (print) 20240456807 | Canadiana (ebook) 20240456947 | ISBN 9781487550349 (softcover) | ISBN 9781487550325 (hardcover) | ISBN 9781487550370 (EPUB) | ISBN 9781487550356 (PDF)
Subjects: LCSH: Shipbuilding industry – Poland – Case studies. | LCSH: Shipbuilding industry – Croatia – Case studies. | LCSH: Shipyards – Economic aspects – Poland – Case studies. | LCSH: Shipyards – Economic aspects – Croatia – Case studies. | LCSH: Poland – Economic conditions. | LCSH: Croatia – Economic conditions. | LCSH: Poland – Politics and government – 1989– | LCSH: Croatia – Politics and government – 1990–
Classification: LCC VM299.7.P7 B7813 2025 | DDC 338.4'762382009438 – dc23

___

Cover design: Val Cooke
Cover image: Alexey Pushkin/Shutterstock.com

The German and European Studies series is funded by the DAAD with funds from the German Federal Foreign Office

 Deutscher Akademischer Austauschdienst
German Academic Exchange Service

We wish to acknowledge the land on which the University of Toronto Press operates. This land is the traditional territory of the Wendat, the Anishnaabeg, the Haudenosaunee, the Métis, and the Mississaugas of the Credit First Nation.

University of Toronto Press acknowledges the financial support of the Government of Canada, the Canada Council for the Arts, and the Ontario Arts Council, an agency of the Government of Ontario, for its publishing activities.

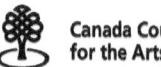 

Contents

List of Maps vii

List of Illustrations ix

Acknowledgments xi

1 Weathering the Storms of Transformation: Shipbuilding and Social Change in Eastern Europe and the EU since the 1970s 3
   The Long and Great Transformation 8
   Polanyi versus Fukuyama 11
   The Paradoxes of Privatization 17
   Comparative and Global Business History 23
   Social Advancement and Differentiation 28
   Methodology and Sources 35

2 Forever on the Verge of Going Under: A Tale of Two Shipyards 43
   Global Economic Cycles and the Late Socialist Transformation 49
   The Crisis and Failed Solutions of the 1970s 55
   On the Brink of Financial Ruin 65
   Continuing to Muddle Through in the Post-Socialist World 70
   In with the EU, Out with the Shipyards 79
   Afterlife(s) 88

3 A Safe Haven? The Role of the State in the Transformation 92
   The Shipyard's Porous Boundaries 94
   The Meandering Leviathan 102
   Rhetorics of Legitimation and Mechanisms of Social Protection 111
   When Debt Management Becomes Blame Management 119

4 Welded Together: Community Building in the Shipyards 124
   The Subculture of Uljanik and Stocznia Gdynia as a Food and Leisure Cooperative 129
   The Multifunctionality of the Shipyard: Housing, Sports, and Welfare 134
   Coping with Collapse: Shipbuilding, Hierarchies, and a Longing for Community 144
   Fragile Communities: A Metaphorical Approximation 155

5 Added Value: Ships, Labour, and the Production of Meaning 161
   The Reduced Value of Industrial Labour: From Communism to Post-Socialism 164
   The Work Ethic and Solidarity of the Past 170
   Rifts and Compromises: New Social Values and Economic Practices 180
   New Horizons on the Old Industrial Landscapes 185

6 Keel Up? The Future of Shipbuilding in the EU 192
   China and the Situatedness of the Transformation 194
   The Hidden Liveability of Transformation 197
   The Temporality of Transformation 201
   After Neoliberal Hegemony 209

Postscript 213

Notes 217

Bibliography 241

Index 261

Maps

1.1 Localization of the two case studies 4
2.1 Outline map of Uljanik shipyard, 1990 44
2.2 Outline map of Gdynia Shipyard, 1995 47
4.1 City plan of Pula indicating apartment blocks and houses with company flats, 1991 135

Illustrations

1.1 Uljanik shipyard in the Bay of Pula, 1962   6
1.2 Global market shares of major shipbuilding regions, 1970–2020   15
1.3 March by Gdynia Shipyard workers to commemorate "Black Thursday," 17 December 1989   32
1.4 Cover page of Uljanik company newspaper, January 1986   39
2.1 Ship engine workshop in Uljanik, mid-1950s   51
2.2 New dry dock in Gdynia, opened in 1976   56
2.3 *Berge Istra*   56
2.4 *Crystal Ray*   74
3.1 *Uncertainty to the Very End*, from Uljanik company newspaper, 1982   104
3.2 Two workers reading a newspaper report about Edward Gierek's appointment as first secretary of the Polish United Workers' Party, 1970   113
4.1 Rock Club Uljanik, 2021   131
4.2 Components of the "social standard" depicted in *Uljanik*, 1986   138
4.3 Uljanik's electricians and their cleaning lady, late 1970s   151
4.4 Ship launch ceremony in Gdynia, 1970s   157
5.1 Welding work being carried out on the hull of the *Bailadila* supertanker in Uljanik, 1971   165
5.2 Uljanik workers receiving certificates for twenty-five years' service, December 1986   176
5.3 Celebration after the twenty-five years' service ceremony in a restaurant, Pula, December 1986   176
5.4 Welding work in Gdynia, 1973   178
5.5 Crist shipyard, 2020   189
6.1 The dredger *Ibn Battuta* in the shipyard in Pula, 2021   212

Acknowledgments

One eminent need expressed by the people we interviewed in Gdynia and Pula, across generations and professions, was that of community. Community also shaped the creation of this book: it is the result of a collective effort by six authors who have jointly researched – each from a different perspective – the fate of the two shipyards from the 1970s to the present day. In the spirit of multi-disciplinarity we used a variety of methods, ranging from historical archive-based document analysis to oral history and ethnographic fieldwork. The book was written collectively; everyone is responsible for everything, regardless of who wrote the first lines in which chapter. In this respect, this book is not only a contribution to the study of the transformation period but also a call for transformative scholarship along multi-perspectival and interdisciplinary lines.

The book is the sole responsibility of its authors, but they could hardly research and write without outside support. First and foremost, we would like to thank the former employees of the two shipyards, and the experts and decision-makers who shared their knowledge and memories with us. Special thanks go to Radivoj Jelenić and his helpful staff at the Uljanik shipyard. The head of the documentation department granted us easy access to the company archive and provided us with a desk in his office where we could read the documents. We would also like to thank everybody who supported us in the research process and later in the preparation of the manuscript: Iwona Kochanowska in the press archive of the newspaper *Polityka*; Mateo Dragičević, Ana Rebić, Sara Žerić, and Davor Zufić for newspaper analyses in Pula; Theodora-Tiha Loos and Fatima Ajanović for transcribing interviews; and Johannes Nüßer, Anita Biricz, and Amina Smajlović for completing and correcting the notes and the bibliography. We especially thank Iris Engemann and her company Berlin-Text for research support and their help in editing. Irena Remestwenski, the manager of the Research Centre for the History of Transformations at the University of Vienna (RECET), provided us with organizational support – we wish to thank her as well. We would also like to thank Carla Welch for her sensitive and creative translation.

Our academic partners in the research locations provided helpful information about sources, arranged contacts, and gave us precious feedback. Thanks therefore especially to Igor Duda, Andrea Matošević, and Igor Stanić from the University of Pula and Tajana Ujčić from the Historical Museum in Rovinj, and to Jacek Friedrich and Łukasz Jasiński from the Museum of the City of Gdynia. Thanks are due to the Historical and Maritime Museum of Istria in Pula, and in particular to the curator Katarina Marić, the head of the photo collection Lana Skuljan Bilić, and the director Gracijano Kešac, for organizing the digitization of the Uljanik shipyard's magazines – in good time, before the shipyard went bankrupt, when Uljanik's company archive faced an uncertain future – and for providing photographs from the museum's photo archive. Darija Hofgräff Marić from the Croatian State Archive in Zagreb unbureaucratically obtained digital copies of images from the shipyard's history for us. We thank Sara Žerić for sharing private photos of her grandfather, who spent his professional life at Uljanik. Ewa Konkel and Jacek Kołtan from the European Solidarity Centre in Gdańsk gave us access to a large collection of images of Gdynia Shipyard and allowed us to use some of the photographs for this publication. We benefitted greatly from contact with the indefatigable Kalman Žiha, the long-time editor of the Croatian shipbuilding magazine *Brodogradnja*; he gave us easy access to back issues of the magazine, and organized a project presentation for the engineers and experts of the shipbuilding faculty in Zagreb, which was very instructive for us. Patricija Softi, a journalist from Pula, gave us insights into the events surrounding the bankruptcy of the company. Ratko Radošević, a long-time employee at the newspaper *Glas Istre*, and Vladimir Sinčić, then managing director of Uljanik's pensioners' club (Klub umirovljenika Uljanika), greatly helped us by establishing contacts with interview partners.

From the preparation of the project proposal to the completion of the manuscript, we received critical feedback from numerous colleagues; without aiming for completeness, we would like to especially thank: John Connelly, Elizabeth Dunn, Valentina Fava, Alison Frank, Dierk Hoffmann, Sarah Graber Majchrzak, Dimitra Kofti, Pavel Kolář, Claudia Kraft, Carolin Leutloff-Grandits, Hugh Murphy, Rainer Liedtke, Vjeran Pavlaković, Jeff Pennigton, Tanja Petrović, Sanja Puljar D'Alessio, Jure Ramšak, Sabine Rutar, Susanne Schattenberg, Reana Senjković, Mark

Spoerer, Annemarie Steidl, Jochen Tholen, Pieter Troch, Marcel van der Linden, and the two anonymous reviewers of the English manuscript.

As the readers will see, our shipyards needed public support; it is no different for research projects like this one. So, last but not least, we thank the two main institutional sponsors of our project "Transformations from Below: Shipyards and Labour Relations in the Uljanik (Croatia) and Gdynia (Poland) Shipyards since the 1980s": the German Research Foundation (Deutsche Forschungsgemeinschaft, DFG) and the Austrian Science Fund (Fonds zur Förderung der wissenschaftlichen Forschung). The bilateral project was completed from 2016 to 2020 at the Leibniz Institute for East and Southeast European Studies in Regensburg and the Institute of Eastern European History and RECET at the University of Vienna. We also thank our home institutions, the University of Vienna and IOS Regensburg, as well as the Leibniz ScienceCampus "Europe and America in the Modern World" in Regensburg for additional financial support for this publication. The DFG-funded Center for Advanced Studies "Universalism and Particularism in European Contemporary History" at Ludwig-Maximilians-University Munich provided one of the authors with the time needed to work on the English translation of our manuscript.

Those who think a project on shipyards would be out of place in Regensburg and Vienna are mistaken: both cities (for Vienna, the suburb of Korneuburg to be precise) were important shipyard locations, and Korneuburg even built ocean-going vessels. The shipyard in Regensburg fell victim to the wave of bankruptcies in the German shipbuilding industry in the mid-1970s, while the one in Korneuburg did not survive the crisis of state-owned industry in Austria in the 1980s. All the more reason, then, to write the contemporary history of European industry beyond the East-West dichotomy.

IN THE STORMS OF TRANSFORMATION

1 Weathering the Storms of Transformation: Shipbuilding and Social Change in Eastern Europe and the EU since the 1970s

The Polish port city of Gdynia (see map 1.1) shows the sunny side of post-communist transformation. The town is bathed in white and blue, with its countless seagulls soaring through the sky, hundreds of sailboats anchored in the marina, and the bright white deck shoes of the contented boat owners. Even the water of the Baltic Sea is no longer murky and foul as it was under state socialism, but instead a pale blue and – due to climate change – noticeably warmer than it used to be. Gdynia's old port is hardly used for transportation anymore, instead primarily serving consumers. The city has something for everyone, whatever their taste and budget – from French fries to upscale Mediterranean cuisine, from cheap baseball caps to smart white linen suits, from inexpensive private rooms in the hinterland to a suite on the thirty-eighth floor of the "Sea Towers" skyscraper. Even the nearby Ferris wheel is dwarfed by this 460-foot apartment building, though the former still affords a perfect view over the shipyards and the port previously used by the Solidarność trade union. The port of Gdynia played a crucial role for Solidarność in 1980/1, enabling it to smuggle the illegal printing and copying equipment needed to produce the newspapers and pamphlets that would mobilize supporters throughout the country.

The cranes and the dockyards, however, have now been relegated to the background. Whole freight ships and tankers are no longer built in Gdynia. Since the global financial and economic crisis of 2009 and the resulting bankruptcy of the major shipyard Stocznia Gdynia S.A.,[1] shipbuilding activities have been limited to smaller enterprises and special orders, repairs, and the construction of components such as command bridges. This can be a profitable business, but it feels much less spectacular than building huge ships entirely from scratch. The remaining shipbuilding companies in Gdynia manufacture only niche products

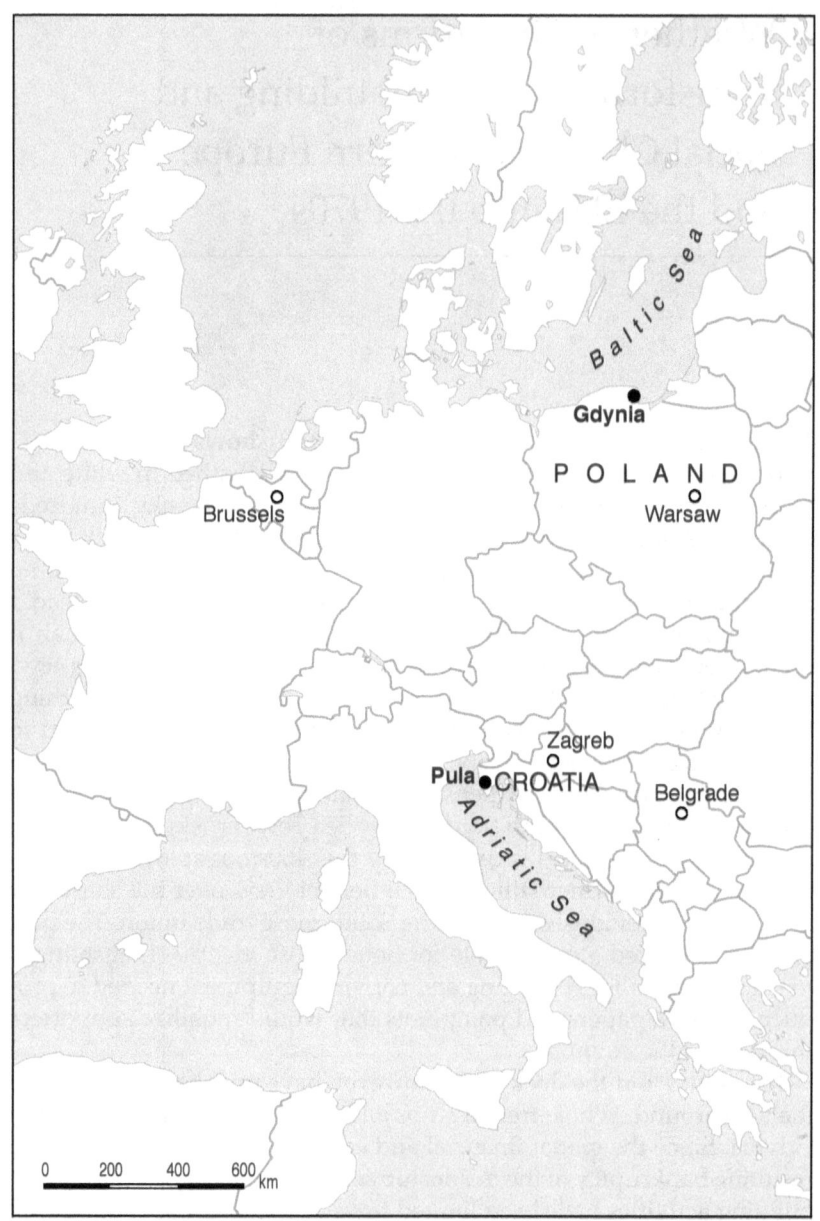

Map 1.1. Localization of the two case studies
Source: © J. Zwick, Ingenieurbüro für Kartographie. Base map courtesy of Natural Earth.

and modules. The local and regional economy is no longer characterized by heavy industry, epitomized by "coal and steel" – the paradigm of the *trentes glorieuses* in Western Europe as well as in the countries of state socialism.[2]

The rise of tourism and the service industry can be viewed as a successful structural transformation – and indeed, Gdynia is better off than places like Bremen, Rostock, or large parts of the Ruhr region in Germany in terms of economic indicators such as growth, unemployment, or GDP per capita.[3] But the absence of the towering, nearly 1,000-foot-long ships that secured Gdynia Shipyard a place among the world's ten largest ship manufacturers in 2004 is felt in a broader sense as well. These ships were powerful symbols of industrial modernity, honoured with rituals marking their life cycle: their launch, christening, and maiden voyage in the inner harbour were huge public events that reflected and strengthened the connection between the shipyard and the urban community. They were part of day-to-day life in the Polish People's Republic, when the shipyard was named after the Paris Commune, as well as during the following two decades, when Gdynia and the entire country were being transformed to a market economy. Large merchant vessels stood for Gdynia's integration into a modern world bound together by means of seaborne transport and international trade. Yet, despite brisk business in the late 1990s and its proud history, Gdynia Shipyard was not spared the fate of so many other European shipyards. It proved unable to withstand East Asian competition and the withdrawal of state support after Poland's accession to the EU, and consequently went out of business in 2009.

The upshot of the post-socialist economic transformation, which occurred more rapidly and impacted more levels of society than the structural shift to a post-industrial society in Western Europe, is clearly visible in the streets of Gdynia. Though still home to around 250,000 inhabitants, the former workers' city has become a tourist metropolis. Nowadays, you have to venture deep into the industrial area around the dry dock, built in 1976, to find a worker in a boilersuit, and you are unlikely to see any women at all. Female industry workers were no longer in demand following the abrogation of the socialist employment policies. As a result, all the more women work for the many service enterprises that have proliferated in Gdynia in the past three decades. Does all this mean that the transition from an industrial to a service society in Poland and post-communist Europe occurred as if in a time-lapse? Was it merely a question of catching up with the West?

The specific dynamics and ambivalences of the post-socialist transformation and its consequences for the entire EU can be better understood

6   In the Storms of Transformation

Figure 1.1. Uljanik shipyard in the Bay of Pula, 1962
Source: Courtesy of Croatian State Archive, AGEFOTO Fund, HR-had-1422.

by examining more than one location and country. The Croatian city of Pula (see map 1.1), in the south of the Istrian peninsula, is also home to the abandoned cranes of a once grand and busy shipyard. Although they are now colourfully illuminated at night as a new tourist attraction, they have not been used for their original purpose since early 2018, when the Uljanik shipyard drifted into insolvency. There is no need for a Ferris wheel or a room on the top floor of a luxury hotel to catch a glimpse of Pula's shipyard and docks (see figure 1.1), however: these can be found immediately behind the well-preserved Roman amphitheatre and Temple of Augustus that dominate the centre of the old town. In Pula, too, shipbuilding has a long history, with the docks founded in 1856 under Habsburg rule. But neither its history nor its key importance for Croatia's trade balance could prevent the shipyard in Pula from suffering the same fate as its counterpart in Gdynia. Even the protests of the many workers, male and female alike, whom this book will lend a voice were of no avail.

The nearly identical time lapse between the bankruptcy of the two shipyards (2009 and 2018, respectively) and their respective countries' accessions to the EU in 2004 and 2013 is no coincidence. In both cases, changes to the fundamental business conditions brought about by the strict EU competition laws meant that the business model these enterprises had once used, which saw them being kept afloat by government aid during frequent financial emergencies or even for extended periods, was no longer tenable. The EU brought not only access to the common market, but with it a completely new regulatory regime, which pulled the rug out from under the state in its role as the shipyards' insurer of last resort. In 2009, Poland was ordered by the European Commission to recoup the billions of złotys it had paid out in state aid to Gdynia Shipyard, and in 2018, the Croatian government was forbidden from granting new subsidies to Uljanik to plug a massive hole in its balance sheet. Both decisions symbolized the end of a long practice of muddling through with a helping hand lent by the state, and pushed the shipyards into bankruptcy. Does this mean that the socialist modernity based on heavy industry did not end until the countries of Eastern Europe joined the single European market? This is one of the questions that will be discussed in this book.

For Croatia, the failure of the shipyard in Pula (along with the one in Rijeka that was part of the same company) represented a far worse economic blow than the loss of the Gdynia works did for Poland. In its heyday, shipbuilding contributed a considerable share of Croatia's total industrial production and generated more than one-fifth of its export revenues. Also lost were the majority of loans and guarantees granted to the shipbuilding industry by state-owned or government-related institutions, which in 2010 alone amounted to around 4 per cent of Croatia's GDP.[4] The losses in the industry thus even exceeded what some other countries spent on "saving" their banks in the wake of the 2008/9 financial crisis.

Nevertheless, the end of the old shipyard is less noticeable in Pula's everyday life than many inhabitants had feared. True to the old motto "When Uljanik catches a cold, the entire city starts coughing," the shipyard's managers had predicted a social catastrophe in the event of its bankruptcy (as a convenient justification for their request for state help). But in fact, only 2,000 workers were employed at the company shortly before it closed, a fraction of its peak workforce of more than 8,000 in the late 1980s. As in Gdynia, the 2,700 apartments previously owned by Uljanik had already been sold in the 1990s, along with a considerable number of holiday homes and leisure facilities. As a result, relatively few people were immediately affected by cessation of production. When the shipyard was in its death throes – its management was forced to announce massive losses in late 2017 – many engineers and specialized workers found new jobs elsewhere, particularly with Italian and

German shipbuilders and other industrial enterprises. Besides, tourism had long since supplanted the construction of ships as the leading economic factor in Pula – and in fact in Croatia as a whole. As a result, many owners of the former company apartments or of houses in the countryside financed with cheap company loans during socialist times are now making decent revenues providing tourist accommodation.

It also explains why Uljanik's impending closure sparked only short-lived protests on the part of its workers. The lack of mass demonstrations is another indicator of the fundamental change that ultimately resulted from Croatia's accession to the EU: shutting down a major industrial enterprise because the state refused to continue its support became politically feasible. The situation in Gdynia ten years earlier had been much the same, with the (former) members of Solidarność staying at home when the liberal-conservative coalition announced the closure of the country's last major shipyard. After the end of state socialism, the shipyards gradually lost their economic and social significance, and even more so their symbolic power as emblems of industrial modernity and progress – and thereby their ability to depict their concerns as issues pertaining to society as a whole. For the respective governments, this in turn lowered the political cost of denying the shipyards further state subsidies, especially since they were now also able to blame the European Commission in Brussels.

The Long and Great Transformation

The years 1976, 1981, 1989, 2001, 2004, and 2009 mentioned above – to which we can add, for Croatia, the Yugoslav constitutional reform of 1974, the collapse of Yugoslavia in 1991 and subsequent war lasting until 1995, and the country's accession to the EU in 2013 – suggest that the period of transformation was divided into phases and included multiple sharp breaks. Transformation obviously took much longer than the political regime change alone. This relativizes the break of 1989, which transitologists and experts on the post-communist transformation have subsequently often viewed as a zero hour.

This traditional interpretation is contravened not least by the emergence and growth of niches for private business under state socialism, which enabled a form of capitalism "from below" in the 1980s even before the reform politicians proclaimed it as the promise for the future. The growing willingness to implement radical reforms – or at least an awareness of the need for them –, along with the integration of Eastern Europe into the global economy starting as early as the 1960s, suggest that the chronology of the transformation began long before the communists lost power. Limiting the upheaval to a single year – the much vaunted *"annus mirabilis"* 1989 – inappropriately condenses what was a much longer process. While the regime change certainly culminated in the collapse of the Soviet Union and the dissolution of Yugoslavia in 1991, the beginnings of transformation lay much further back.

When, then, did the transformation begin? The deep economic crisis and the mounting social protest point towards the 1980s as a period of condensed changes, as a "saddle period" (*Sattelzeit, pace* Reinhart Kosselleck) of sorts that marked the shift to a post-industrial, increasingly neoliberal capitalism in the West, while the contradictions inherent in the systems of state socialism became untenable. As our two case studies illustrated, this was especially visible at the business level. It could even be argued that Poland and Yugoslavia, and hence two variants of socialism, went bankrupt with their shipbuilding industries. The 1980s thus marks the "beginning of the end,"[5] which was nevertheless just as unforeseeable as the "end of history" in 1989.

Following the first oil price shock in 1974, specific developments in the shipbuilding industry, which we will outline in chapter 2, came together with political decisions and a shifting international economic landscape to set in motion a number of long-term structural changes. The socialist economies ultimately lacked the necessary flexibility to adapt to these changes. As Fritz Bartel has shown, they especially lacked the political legitimacy needed to undertake measures to develop a new, more productive economic model and to impose welfare cuts. Instead they embarked on a course of increasing indebtedness.[6] Shipbuilding illustrates the significance of this shock better than many other industries, since it was directly affected by it. As a consequence of skyrocketing oil prices and the associated decline in global trade, the international market for ships fell into a persistent slump that would last until the 1990s.[7] Shipyards, including those in socialist economies, also faced specific financing needs because of the long production cycles of ships, which made them particularly sensitive to changing interest rates.

For the Western economies, the experience of the oil price shock created a strong impetus for a structural shift towards the service and IT sectors.[8] It also facilitated the emergence of austerity and neoliberalism as new paradigms of economic policymaking in the 1980s.[9] The socialist countries, by contrast, continued investing into the expansion of heavy industry – not least owing to their long planning cycles. The erection of the dry dock in Gdynia in 1976 was one of Poland's biggest investments – financed with Western loans.[10] Because of this modernization attempt, Gdynia Shipyard, named after the Paris Commune in 1951, surpassed the economic significance of its counterparts in Szczecin and

Gdańsk. The Lenin shipyard in Gdańsk was better known internationally for political reasons, but the shipyard in Gdynia is in many ways a better *pars pro toto* for the economic development of Poland and Eastern Europe.[11]

Trouble had also been brewing in Yugoslavia since the mid-1970s. In 1976, the Yugoslav government enacted the "Law on Associated Labour" that stipulated a radical decentralization of enterprises. Firms were dissolved as legal entities and divided into what were known as "basic organizations of associated labour." Although this transformation was supposed to strengthen self-management, it actually led to further bureaucratization and effectively paralysed management.[12] This in turn exacerbated the crisis in the Yugoslav shipbuilding industry – as well as other industry sectors – because it made it even more difficult for them to adapt to the new international conditions. Yugoslav shipyards, all but one of which were located in the Socialist Republic of Croatia, produced primarily for export and faced global competition. During the 1980s, the country's economic woes were rapidly spiralling into what ultimately proved to be a fatal crisis with high inflation, rising unemployment, rampant debt, and a lack of liquidity. The Uljanik shipyard shows these developments as if under a magnifying glass.

Meanwhile, Poland also came to realize that socialist "modernization" supported by capital imports was aggravating the country's economic problems rather than solving them. The year 1976 saw the first large-scale strikes along with the establishment of the Workers' Defence Committee (KOR), a precursor of Solidarność. A major recession – which theoretically should not even have been possible in a planned economy – began in 1979, forcing the regime to allow the founding of Solidarność in 1980 as a political pressure valve of sorts. This is why the year 1980 marked a turning point for late socialism in Poland. In Yugoslavia, however, 1980 was pivotal because Tito, the charismatic ultimate arbitrator of political disputes and figurehead of the Non-aligned Movement, died that year. The second oil price shock in 1979 and the ensuing worldwide economic decline added further pressure, with the Yugoslav shipyards feeling the effects of the global decrease in demand for transport ships even more than their Polish counterparts, since they had relied almost entirely on exports. The year 1980 was thus the culmination of the political and economic contradictions of late socialism, which amplified inherent inconsistencies and accelerated tectonic changes. In hindsight, we can thus identify 1980 as a turning point in the history our "long" transformation.

The two variants of socialism explored in this book – state socialism and Yugoslavia's self-management – came to an end in Europe between

1989 and 1991. Soon after this, the systematic decline of social democracy – which may at some point in the future be considered the third, civic variant of socialism – accelerated in the West. But the individual and group-related attitudes and behaviours of the blue- and white-collar workers as well as the management of the shipyards obviously did not change within just two years. Incidentally, the same applies to the West and the structural changes initiated there following the first and the second oil crisis. There is evidence of this in management practices as well as in day-to-day work – in other words, in the corporate history as well as the history of labour examined during our research project.

Polanyi versus Fukuyama

Besides being temporally and regionally confined, the term "transformation" as used after 1989 contained a double teleology. When the dictatorships in southern Europe and Latin America came under pressure during the 1980s, eventually being supplanted by democracies, political scientists initially employed the Spanish term *"transición."* Working as advisers to the first post-communist government in Poland in 1990, economists Jeffrey Sachs and David Lipton further developed this concept into what they called the "dual transition," which included the conversion from planned to market economy alongside the political change from dictatorship to democracy.[13] Similar to Francis Fukuyama's theory of the "end of history," Sachs's and Lipton's basic assumption was that the establishment of liberal democracy and a free market economy were intrinsically linked. However, the term "transition" soon attracted criticism due to its teleological nature and normative underpinning, as well as the fact that the actual developments turned out to be far more ambivalent.[14]

In order to capture these inconsistencies and differences as well as the further-reaching societal and political dimensions of the process, sociologists introduced the term *transformation*. The intention was to produce a concept that was more process-oriented and less prescriptive. The consolidation of democracy and the universal application of the free market principle, especially privatization, nevertheless formed the foundation for both research terms. Even proponents of the transformation paradigm assumed that something genuinely new was developing "after" the end of state socialism – they simply did not take the unambiguously positive view that the transitologists did. Rather, criticism of unbridled capitalism and the loss of the welfare state components of state socialism is a key tendency in transformation research.[15]

The temporal delimitation from state socialism and the notion of a fundamentally different world emerging from its ruins are constitutive for this area of research. Using our case studies, however, we argue that there was in fact much continuity. Take, for example, the role of foreign capital. Both in Poland and Yugoslavia in the 1970s, communist governments hoped to achieve comprehensive economic modernization with the help of capital imports from the West – and a similar hope was evident in the 1990s and 2000s in the form of foreign direct investments and subsidies from the EU. Similarly, (re)integration into the world market was a goal both before and after the dramatic upheaval, and this is particularly apparent in the shipbuilding industry. As chapter 3 will show, practices of state aid for industry is another area of significant continuity, and in chapter 4 we will discuss the continuation of enterprise-based welfare and community structures across the divide of 1989.

Our understanding of transformation is, therefore, inspired by an older concept: Karl Polanyi's *Great Transformation*, a work that has been attracting renewed attention since the turn of the millennium owing to the frequent crises of global capitalism. It is especially well suited to the interpretation of the post-1989 changes in Eastern Europe. Polanyi's opus magnum primarily uses the example of England to describe the transformation from an agrarian to an industrial society, the rise of global free trade and of laissez-faire capitalism, along with its social and political consequences, from the late eighteenth century to the interwar period.[16] While we do not adopt a strictly Polanyian perspective in the way that various academic publications on the "varieties of capitalism" have done very productively,[17] *The Great Transformation* is a source of inspiration for us in at least three regards. First, it suggests an extended temporal dimension that does not link transformation to political breaks but instead to shifts in the structures of the political economy and of societal organization. Second, Polanyi adopts a sociologically and anthropologically oriented perspective, emphasizing that changes in the spheres of meaning produced by social groups are a key aspect of the "great transformation." Transformation does not simply wash over people; rather, it is made and remade by them – though of course they are often not aware of it. This is why we called the project from which this book has emerged "Transformation from Below."[18]

Third, Polanyi shows that a profound transformation of the socio-economic and everyday environment like the one that followed the collapse of state socialism has a political and social price. The social fault lines and frictions appearing in the framework of global laissez-faire capitalism – as the neoliberal order that has been emerging since the

1980s might be labelled with Polanyi in mind – and the abandonment of an "embedded capitalism" produced a "social need for protection." Polanyi's work does not answer the question as to what political currents this desire for protection may encourage. Especially in Poland, the radically anti-communist wing of Solidarność challenged liberal democracy.[19] In the eyes of the workers supporting the Law and Justice (PiS) party, which governed in 2005–7 and 2015–23, the closing of the shipyards in Gdynia, Gdańsk, and Szczecin was a conspiracy by the EU and Angela Merkel to remove the Polish competition for the German and other Western European shipyards by prohibiting state subsidies. The last PiS-led government dreamt of, and partially created, a paternalistic welfare state interfering in the economy at will in the name of "national interests" and resurrecting the pride of the naval nation. This constituted a countermovement as predicted by Polanyi.

From an abstract point of view, the closure of the shipyards resulted from a conflict between different levels of public administration and their respective relations to the economy. It was also about the nation state's power to intervene, as well as the expectations of society. On the one side stood the Polish government, which pumped further subsidies worth around 1.43 billion euros into the Gdynia shipbuilding group between 2004 and 2007.[20] And the Croatian government did much the same until the country joined the EU in 2013, waiving roughly 1.3 billion euros of debt owed by the Croatian shipyards and allocating a further 500 million in restructuring funds. On the other side was the European Commission, which views itself as a guardian of free competition and therefore imposes strict regulations on state subsidies, allowing them only in exceptional situations and with a clear restructuring plan attached. In the case of Poland, in 2009, the EU demanded the repayment of the state aid granted, which caused the immediate bankruptcy of the shipyard; in Croatia, the prohibition of further subsidies in 2018 led to the same result. All this confirms Polanyi's assessment that free market capitalism does not emerge as a kind of natural order, but is implemented by the state – in this case by the EU, which wields state-like powers in economic matters.

The two levels of state policy, the EU and the nation states, obviously differed in their understanding of the economy and the role of the state within it. This is not least due to their different political foundations: while the government of a nation state is directly accountable to its electorate, nobody can vote out the European Commission. The latter primarily sees itself as a warden of the laws of the EU and not as an enforcer of the people's will. As Quinn Slobodian has shown, the primary concern of the intellectual pioneers of neoliberalism was not austerity but to protect the free market from the constraints of democracy.[21] The democratic control of the national government explains the deviation from neoliberal dogmas already discernible in Poland during the 1990s – despite copious rhetoric to the contrary – under Minister of Economic Affairs Balcerowicz.[22] In Croatia, a radical market-oriented policy never really took hold. The national conservatives and social democrats, who have taken turns governing the country since 1991, continually promised state intervention to attenuate the social costs of transition – and the population expects nothing less. Modifying Polanyi's terminology, we might speak of the need for *political* protection, which was in fact provided and safeguarded the shipbuilding industry in Poland and Croatia from deeper cuts and closures for a considerable time. Yet, as the results show, economic protectionism did not save the shipyards in the long term, and was not even supposed to exist in the enlarged EU. It would be inaccurate to label this a purely Eastern European problem, however. Over the past fifty years, Europe's share in the global production of ships has dropped from more than 80 to below 10 per cent (see figure 1.2).

As suggested by Polanyi, social protection was easier to achieve than maintaining global competitiveness in a market that was not a level playing field because Europe's East Asian competitors, who gained global dominance in the same period, enjoyed the continuous backing of their governments. At the same time, the EU supported new industrial developments beyond the "old heavy industry," thus facilitating the move beyond "coal and steel" that the West had begun decades earlier.

Many local stakeholders interpreted the bankruptcy of the shipyards as part of a widespread process of deindustrialization accompanied by a decline in prosperity. There is a powerful trope that Eastern Europe's industrial economies have in fact declined since the end of state socialism. The in many ways groundbreaking ethnographic accounts of the fate of industrial workers in post-socialism often engender the impression that there is nothing left of the former industrial landscape of Eastern Europe. With this book, we seek to introduce some nuance. Based on the share of industry in overall employment numbers and the gross domestic product, most of the formerly socialist countries depend more heavily on industrial production today than the original regions of the Industrial Revolution. In 2021, industry contributed almost 28 per cent to the GDP of Poland, which was more than Germany's 26.7 per cent. Croatia, which was never as industrialized as Poland, had a lower share (19.8 per cent), but this was still substantially higher than the once pre-eminent industrial centres of the world, the United States and the

Figure 1.2. Global market shares of major shipbuilding regions, 1970–2020 (in Mio GT)

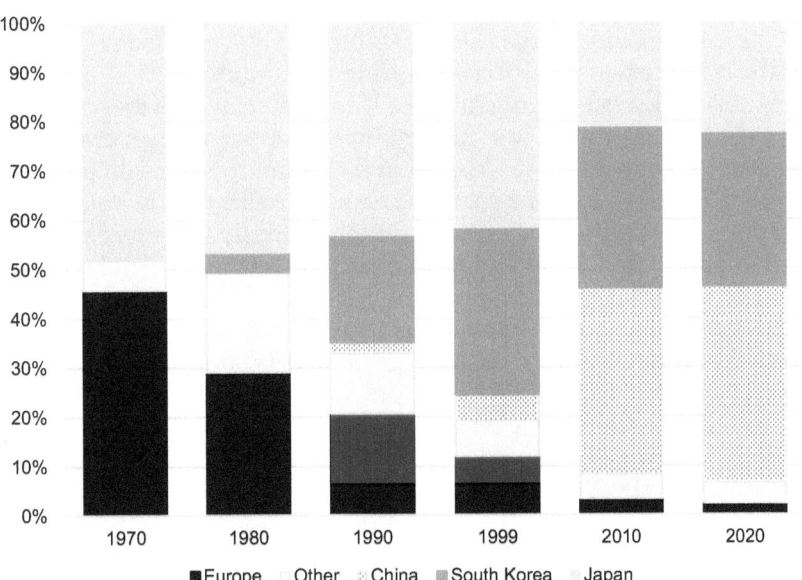

Sources: Lloyd's Register of Shipping, *World Fleet Statistics*, 1981, 1991, 1999, accessed 29 March 2024, https://hec.lrfoundation.org.uk/archive-library/world-fleet-statistics; Shipbuilder's Association of Japan, "Shipbuilding Statistics," March 2019, https://www.sajn.or.jp/files/view/articles_doc/src/addd344c6169359e87a5f39d64e02173.pdf; and Shipbuilder's Association of Japan, "Shipbuilding Statistics," September 2021, https://www.sajn.or.jp/files/view/articles_doc/src/dec775d2e4188a22f74082282a715f20.pdf.

United Kingdom (17.9 and 17.5 per cent, respectively). It is notable that in Poland, industrial decline came to an end in 2000. Since then, the importance of manufacturing for the Polish economy has risen slightly again. As an aside, in the People's Republic of China, industry produces 39.4 per cent of value added. In terms of employment, the picture is similar: 32 per cent of all employment in Poland and 28 per cent in Croatia is in manufacturing, while the respective values for Germany, the United States, and the UK are 27, 20, and 18 per cent.[23] Our two case studies provide a better understanding of the dynamics of deindustrialization as well as re-industrialization, illustrating that not all industrial enterprises disappeared – but those that did were often the ones whose existence was closely tied to the structures of the socialist state and which were unable to attract foreign investors. But even these enterprises have often given way to new, smaller, and more flexible industrial firms. The former communist countries, thus, experienced both "creative" (*pace* Schumpeter) and destructive destruction.

The closure of industrial companies produced a great many "losers" of the transformation, especially during the turbulent 1990s, not least because of the wide-ranging role these firms had played in the provision of welfare facilities under socialism. But the notion of a general social decline, which is quite widespread in social anthropological literature, is as erroneous as that of a complete deindustrialization. In Poland, for example, prosperity rose quickly and continuously once the economic slump of the early 1990s had been overcome, and not even the global recession of 2008/9 could stop this trend. Poland's GDP per capita adjusted for purchasing power has increased from around one-third to well over two-thirds of the average of all EU countries over the past thirty years. In terms of purchasing power parity, the Polish population today enjoys a slightly higher average wealth than that of Portugal – a country that was considered poor by EU standards in 1990 but at the time was still almost twice as prosperous as Poland. Such a rapid catch-up has rarely occurred in the entire economic history of the world.[24]

Despite this convergence, however, very few people in Poland would refer to the country's development as an "economic miracle." In fact, many people see themselves as "losers" of the transformation and, in reaction to this, now form the electoral base for the right-wing populists, who governed the country in 2005–7 and 2015–23. As Ghodsee and Orenstein highlight, one major reason for the widespread dissatisfaction with the results of transformation is the rapid increase in inequality in societies that prefer a more equal distribution of wealth.[25] This development is linked to the structural changes in the industrial sector and the emergence of a service economy that apparently generates less security and confidence than the period of high modernity after the Second World War – at least for men, who were at the heart of the notions of modernity revolving around industry.

Croatia's overall economic performance provided even more reason for discontent and the sense of loss that is so salient also among the people we interviewed for this book. This was partly a result of the war of independence (1991–5), which not only brought substantial physical destruction but obviously led the government to prioritize military over economic matters. But even beyond the war, Croatia's economic policy was not as dynamic. Nevertheless, Croatia, too, was able to catch up considerably in terms of per-capita GDP. When the war ended in 1995, GDP per capita (adjusted for purchasing power) amounted to 44 per cent of the EU average (or roughly 34 per cent of

German GDP); by 2021, it had increased to 70 per cent of the EU average (Poland stood at 77 per cent, Germany at 120 per cent, and Greece, the poorest of the "old" members at 64 per cent).[26] The comparison with Serbia, which started on a similar level of development to Croatia in 1990 but did not join the EU, is striking: its level of prosperity has fallen behind Croatia's by one-third. Especially along Croatia's Adriatic coast, where the shipyards were located, the tourism boom has considerably improved the standard of living and offers numerous (and oftentimes better paid) alternatives to industrial employment – yet many people still articulate a sense of social degradation. The works of revisionist Marxist Karl Polanyi help us understand how growing prosperity can go hand in hand with alienation from one's own work and the political system rather than increased satisfaction. The close examination of the shipyards in this book likewise seeks to help resolve this supposed cognitive dissonance.[27]

Our reference to Polanyi does not render the concept of transformation as it was understood after 1989 irrelevant. However, we stress that there were many other expectations of democratization and the introduction of a market economy: increasing prosperity, a Europe that would grow together, and the – albeit short-lived – hope of a democratization of enterprises and the economy in general. The latter, which was one of the central promises of Solidarność in 1980/1 as well as the Czechoslovak and East German revolutionaries, rapidly vanished in 1990. The challenges and unexpected difficulties of the shift from the planned to the market economy were so huge, and the acute economic crisis so all-consuming for everyone involved, that there was no time or space left to experiment with new alternatives. The decision by the workers' council of the Pula shipyard to disband and to agree to the conversion of the shipyard into a joint-stock company is emblematic of the fact that the democratization of the political system was accompanied by much less employee participation on the enterprise level. The concentration of power in the hands of the management, too, is an important factor that helps to explain the less than enthusiastic attitudes regarding the outcomes of the transformation.

The Paradoxes of Privatization

In terms of the starting point for their economic reforms, there were parallels between Poland and Yugoslavia that were also shared with other countries in Eastern Europe in the late 1980s: an acute debt crisis and the threat of default – to which the mismanagement of the shipyards had contributed significantly – as well as rising prices that quickly escalated into hyperinflation. In Yugoslavia, unemployment was also skyrocketing. All these factors increased the government's willingness to implement radical economic reforms. The failure of perestroika and its Polish variant (the Wilczek reforms, named after the last communist minister of economic affairs) along with the existential crisis of Yugoslavia, which had been the most important model for a "Third Way" between socialism and capitalism, seemed to leave only one possible approach to economic reforms: liberalization, deregulation, and privatization along with stabilizing measures at the macroeconomic level. This was the so-called Washington Consensus formulated in 1989. One of the immediate consequences of the approach were drastic cuts to government spending. An important goal of these reforms was to attract foreign direct investments to spark economic growth and increase productivity.

The consensus on the necessity of radical reform spanned almost all political camps and currents in Poland in 1988/9: the liberals, anticommunist and church-affiliated conservatives, left-wing adherents of Catholic social doctrine, parts of the left wing of Solidarność close to Jacek Kuroń, and – last but not least – the reform communists. For this reason, and because of its similarities with the Washington Consensus in terms of form and content, we can speak of a "Warsaw Consensus."[28] One of the strengths of neoliberalism has always been its adaptability, and this held true for Poland as well. Far from all of the standard ideological parameters were implemented.[29] At least in terms of actual government policy, neoliberalism was clearly a "soft" ideology that could encompass very many different things.

Yugoslavia, however, lacked a similar internal consensus and international leap of faith after Tito's death. With the end of the Cold War, the multiethnic federation lost its exceptional strategic position between East and West, resulting in a decreased willingness of Western governments to aid the perpetually impecunious country, and the Yugoslav communists no longer resembled a reformist avant-garde compared to the reformers in Hungary or Poland. The Yugoslav variant of socialism, which had not been imposed by an occupying power (i.e., the Soviet Union), was also more embedded in society – to use another of Polanyi's terms – by way of extensive self-management and the legitimacy enjoyed by the communists thanks to their successful liberation war against the Germans during the Second World War. There was therefore no consensus on the need for a radical change in economic policy in 1989. Ante Marković, the last Yugoslav prime minister, initiated drastic austerity measures to stabilize the economy. However, his attempts were thwarted by opposition from individual republics and especially from Serbian party chairman Slobodan Milošević.[30] One of the reasons the opposition to neoliberal policies was so intense in Yugoslavia was that the country had already experienced the conditions imposed by the International Monetary Fund in return for debt relief in the early 1980s. These austerity measures had not resolved the crisis; instead, they exacerbated the economic recession. As a result, Yugoslavia had been suffering from a significant decline in real wages and growing unemployment throughout the 1980s.

Ex post, the rise of Poland – which had been much poorer than Yugoslavia, and Croatia for that matter – is largely attributed to shock therapy.[31] A direct causal relationship between shock therapy and economic growth (and later prosperity) is not discernible, however. Factors beyond the regulative capabilities of economic policy, including the specific timing and geography of reform efforts, had a significant influence on their progress and outcome. Poland had the advantage of being among the trail-blazers of reform towards a market economy, and the West rewarded this with a strong debt reduction. Foreign direct investments arrived earlier in Poland and in greater volume than for the latecomers, which included Croatia due to the war and inhospitable environment for foreign capital. Poland, together with Hungary, came to enjoy European Community support in as early as 1989 through the PHARE program (Poland and Hungary: Assistance for Restructuring their Economies). Another decisive element in the economic recovery of Poland – as well as of Slovenia, Hungary, and the Czech Republic – was the proximity to West European export markets, especially Germany. The short transportation routes and the significantly lower wage level provided strong incentives for transferring the production of industrial goods to these countries. A further factor was the large domestic market promising huge potential for foreign investors – Poland's population exceeds Croatia's by a factor of eight.

The turnaround in Croatia's domestic policy was similarly important for its economic development but with different results than in Poland. Franjo Tuđman's nationalists won the 1990 parliamentary elections against the reform communists, and the newly convened parliament elected Tuđman president of the republic. The new governing party the Croatian Democratic Union (HDZ) was not fundamentally opposed to reforms, but it rejected shock therapy for fear of economic disruptions. It did not want to risk protests resulting from business closures during the war. Instead, the HDZ granted aid to Croatian industry, as evident in the shipbuilding sector. After the end of the war, the government pumped some 300 million euros into the Pula shipyard, for example.[32] The goal of Tuđman's economic policy was to build a national capitalism with Croatian owners – but its result was a half-baked and shady mess of privatizations in which entrepreneurs with ties to the president and his party tended to benefit the most.[33] One of the long-term consequences of this opaque crony capitalism is a fundamental mistrust when it comes to large investors that is still felt in Croatia to this day.

The focus on shock therapy and its alleged success tends to ignore the question of the extent to which this approach was actually applied, even in Poland – its "pioneer." Political scientists Hilary Appel and Mitchell Orenstein referred to the reform rhetoric directed abroad as "competitive signaling," meaning speech acts intended to allure Western investors.[34] There were also considerable differences depending on industry sector and region or location. The greatest "shock" after 1991 was felt by the Polish agricultural and food industries – and thus also the workers in these sectors, especially farmhands.[35] The Polish textile industry largely vanished during the second half of the 1990s and, despite employing tens of thousands of people, its Croatian counterpart did not survive either.[36] Thus, heavily feminized branches of industry were among the first to experience deindustrialization. By contrast, mining, steel, mechanical engineering, and shipbuilding – the very sectors that the communists had built up as key industries, and that were dominated by male workers – initially escaped shock therapy for the most part.[37]

When, in 1991/2, the shipyard in Gdynia was incurring losses of nearly one złoty for every złoty of revenue, the government deferred the collection of unpaid taxes and contributions to social security and pension funds. The following year, 1993, saw similarly extensive debt relief to that which had been granted to the entire country two years earlier. The government, the banks (which were still partly state-controlled), and seven hundred other creditors waived half of the shipyard's liabilities. Croatia took similar steps to prevent its shipbuilding industry from going belly-up. There was a significant difference between this approach and that taken under state socialism, however. Here, the government did not simply pay off the losses with grants as it had done in the 1980s. Instead, it converted a part of the arrears into shares that were then passed on to banks, pension funds, and other enterprises under indirect state control.

By converting debt into shares, in 1993, the Polish government was able to privatize 40 per cent of the shipyard in Gdynia, at least on paper. In 1997, the Ministry of State Treasury sold shares for a further 12 per cent of the company, thus allowing it to claim that the shipyard had been completely privatized. The question of whether this process can be considered a genuine privatization will be answered in chapter 3 of this book. At any rate, the shipyard's management under CEO Janusz

Szlanta, who had started his career in the finance industry,[38] presented the ownership transfer as a great achievement in the sense of "competitive signaling." As it would later turn out, this was not irreversible. In the shipyard's next financial crisis, the state came to its rescue once more, and in 2003, became its majority owner again. Following the logic of the now fully established stock market capitalism, it did this by way of a buy-back of shares, and thus the post-communists who were back in power in 2001 carefully avoided calling this process a renationalization. Szlanta, who had been successful as CEO for a long time, was replaced, and in 2007, the new PiS-led government installed a new director – a lawyer with party connections. The politicization of the shipyard's top management position signalled a regression to a situation familiar from the time of the Polish People's Republic.

The different starting conditions in Croatia meant its privatization efforts followed an even more convoluted path than Poland's. Yugoslavia's self-managed companies formally belonged to their employees, which meant that large enterprises like Uljanik first needed to be nationalized so they could subsequently be privatized and then sold to investors. After nationalization in 1991, privatization was continuously delayed although governments repeatedly claimed it was a priority. First, it was the war that interfered. Then, no international investor could be found. In the new millennium – and after the state had effectively relieved it of the lion's share of its debt – of all the Croatian shipbuilding enterprises, Uljanik seemed the most promising candidate for sale. But the unions resisted acquisition by private investors, and when, in 2009, the global financial and economic crisis hit the shipbuilding industry especially hard, Uljanik became unmarketable. At the same time, the EU was urging Croatia to privatize Uljanik and its other major shipyards. In the absence of international or domestic investors, in 2012, Uljanik's roughly 1.5 million shares were purchased by 7,200 private persons. Most of them were current or former Uljanik employees who believed in the shipyard's future – and suddenly they owned nearly half the company. Much like under Margaret Thatcher in the UK or somewhat later under Václav Klaus in the Czech Republic, a (small) nation of shareholders was created. The ultimate outcome, however, was the return to a similar ownership structure to that of the former Yugoslavia. The company once again belonged to its employees – at least nominally, for the dominant decision-makers were still the state-operated funds with their majority holdings. It goes without saying that these small stockholders had no capital for the necessary investments in the technological modernization of the shipyard.

A major problem that remained unresolved by the transition in ownership was the low productivity of the shipyards in Gdynia and Pula. In a 2008 evaluation of state subsidies in the shipbuilding sector, the European Commission calculated that the Gdynia facility required almost twice the number of working hours as its Western European competitors to build one gross registered ton (GRT, a common unit of measurement for ships). The situation in Pula was similar, if not worse, aggravated by the fact that the shipyard there had no large dry dock, so the bigger hulls had to be manufactured in two parts and welded together. During the 1990s, both companies had been able to make up for their technological deficits thanks to their wage costs, which were much lower than in Western Europe and Japan. This advantage was lost as a result of the revaluation of the Polish złoty prior to Poland's access to the EU and the pegging of the Croatian kuna to the German mark and then the euro. In addition, wages in both countries slowly began to approach Western European levels and were definitely much higher than in China. While this was a welcome development for the workers – convergence with the old EU countries, one of the main goals of the transformation, seemed to be proving more than just an empty promise – it exacerbated the shipyards' problems when it came to costs and competitiveness.

Another important part of the story is, therefore, China's powerful entrance into the world market of commercial vessels in the 1990s (which followed the rapid rise of South Korea from the mid-1970s). Poles and Croatians would not work for the kind of wage the Chinese workers were paid. What is more, most large shipyards in China still enjoy the financial and political support of their government, at least on a regional level. Their success is predicated on a different relationship between the state and the economy than that promoted by the European Union. When economic problems such as a slump in the global demand of ships arise, in China it is not the often-cited "invisible hand" of the market but the helping hand of the state that intervenes – as had been the case in the former Eastern Bloc. In the enlarged EU, however, these two principles competed at the different levels of governance. On the one side were the nation states, which would have liked to act the way they had before 1989 but lacked China's deep pockets, and on the other was the EU, which strictly adhered to the principle of competition. But while the European Union can enforce it competition rules within its boundaries, it cannot do so at the global level. It is in this regard that the history of our shipyards can also shed a light on global developments and their impact on the local level.

Comparative and Global Business History

The fact that the history of the shipyards in Gdynia and Pula was a long and occasionally heart-stopping rollercoaster ride is mostly down to the specifics of the industry itself. Ships are in high demand whenever worldwide trade and the global economy are booming, but during economic crises, the demand slumps equally rapidly. Examples of such slumps are the periods after the two oil price shocks in 1974 and 1979 and after the global financial and economic crisis starting in 2009. During such downturns, orders are often cancelled and prices collapse.[39] The Pula shipyard even founded its own shipping company in the 1980s as a subsidiary to deal with ships that had been built but could not be sold. This construct is reminiscent of the "bad banks" the world was forced to get used to after 2009. In fact, the comparison with the financial world is profoundly relevant in this context, since ships are generally financed in advance via loans. The financializing of shipbuilding further increased its volatility. The shipbuilding crisis in the 1980s, for example, was not only an immediate result of the decline in world trade in the wake of the second oil price shock but also due to the US Federal Reserve raising interest rates, which dramatically increased borrowing costs worldwide.

It is a widespread assumption in general economic history that the countries of the Eastern Bloc did a decent job of protecting their economies against the downward spikes after the oil price shocks, albeit with long-term negative consequences.[40] This does not apply to the shipbuilding industry, however, and especially not Yugoslavia's. During the 1970s, the Croatian shipyards produced 90 per cent of their vessels for the world market. This dependence on global demand took a heavy toll during the two oil crises, when Uljanik lapsed into insolvency for several months due to a number of ordered and manufactured ships ultimately not being purchased by the respective customers. Similarly, a brief period of Gdynia Shipyard being unable to meet its financial obligations in 1984/5 was linked to a large order from a Swedish shipping company, which Gdynia did not fulfil in time. Instead the shipyard had to pay crippling penalties. Venturing into the world market might have brought symbolic capital but was a risky business for socialist enterprises. In addition, both shipyards suffered due to the policy of protectionism pursued by the Western shipbuilding nations during the 1970s and 1980s, which they rightly accused of distorting the market with their generous subsidies.

The crisis in the shipbuilding industry lasted much longer than the global recession in the 1980s. While the worldwide economy recovered quickly from the second oil shock, the demand for ships continued to decline until 1988. A similar situation occurred in 2009, with orders for maritime vessels dropping by more than 70 per cent (measured by tonnage) by 2016. In the end, it seems nobody was able to protect the shipyards in Gdynia and Pula from this downswing and its direct effect on their order books and balance sheets.

The two companies' economic problems were partly self-inflicted, of course. During the 1970s, the shipyard in Pula had relied too heavily on business with Third World countries that were regularly unable to pay for ships they had ordered or hoped to settle their outstanding bills by bartering goods (e.g., bananas, coffee, or cotton in exchange for ships). An alternative customer was found in the Soviet Union, but it, too, did not pay in convertible currency, offering transfer rubles or raw materials instead.

The agony of Comecon and the Soviet Union in the late 1980s pushed both shipyards to the brink of bankruptcy. But under communist rule, shutting them down was politically impossible, as the last communist prime minister of Poland, Mieczysław Rakowski, discovered in 1988. His intention to close the Gdańsk shipyard was met with an indignant outcry – the shipyard's symbolic importance was simply too great. The Polish governments of the 1990s did not dare touch the birthplace of Solidarność either – though whether this political protection ultimately benefitted the shipyards is a different question. A comparative analysis of Poland's three largest shipbuilding enterprises conducted in 2003 revealed that the mismanagement of the Gdańsk shipyard had been by far the worst because its leadership had counted on continuous state support. As a consequence of its persistent losses and lack of prospects, it was eventually taken over by the shipyard in neighbouring Gdynia.[41]

Similarly, in Croatia, the government could not afford to close any of the five big shipyards during the 1991–5 war either, as labour disputes on the home front were highly undesirable. When the Uljanik workers declared a strike in autumn 1992 due to two months of outstanding wages, the company swiftly paid the arrears. Hyperinflation meant the wages fell to an equivalent of 1.25 US dollars per hour by 1994. When the staff subsequently went on strike again, management was once again quick to make concessions and promised a 200 per cent pay raise two days later. The company's precarious situation at the time is also illustrated by the fact that it was temporarily unable to pay its electricity bills, resulting in the power being cut off. After the end of the war, the Croatian government prevented the collapse of the domestic shipbuilding industry with generous financial aid. Although rebuilding the areas destroyed during the war should have been its top priority, Zagreb granted Uljanik a waiver of debt and outstanding social security payments and provided another injection of capital. Poland had effectively done the same in 1993 to give Gdynia Shipyard a fresh start.

In both cases, it is clear that the political stakeholders continued to attribute eminent importance to the shipyards, causing the latter to shy away from implementing radical and painful measures. The governments failed to pursue the economically prudent option of liquidating the enterprises, instead preferring to socialize their losses – for it was of course the Polish and Croatian taxpayers who ultimately had to foot the bill for this priority treatment of the shipbuilding sector.

Chapter 5 looks at the shipyards as places imbued with meaning. It is particularly concerned with the reasons for their gradual loss of significance – for politics as well as for the workers – beginning in the 1990s. As mentioned earlier, to some extent, this was a result of the economic changes – as had been the case two decades earlier in Western Europe, the shipbuilding companies were forced to dramatically reduce their workforce. But even more than this, they lost their symbolic power. Due to the politically desired dichotomy between old and new in the post-socialist countries the shipyards were increasingly regarded as relics of times past instead of emblems of modernity.

For the shipbuilding communities, of course, the shipyards possessed another meaning as well. They were frames of reference that determined numerous aspects of people's everyday life – and even the entire horizon of their existence. This perspective "from below" is pivotal for our fifth and final chapter, which is about the production of meaning. Under state socialism, large enterprises were far more than just sites of production and employment. They offered essential welfare benefits and provided infrastructures for cultural and social activities. Owing to the economy of scarcity, their contribution to basic subsistence was of fundamental importance as well – especially after the imposition of martial law in Poland, when almost all foodstuffs were available only to holders of ration cards. Gdynia Shipyard delivered potatoes and onions by the sackful, and sometimes meat and sweets too. This made the families of workers a privileged community, knit together even more closely by joint holidays and their shared daily working routines.

The goal of concentrating on the "core business," as stipulated in Balcerowicz's reform program and advised by Western consultants, was therefore pursued rather hesitantly by the shipyards. They only slowly sold off their holiday homes and especially the coveted apartments, and these sales were more about obtaining cash for debt repayment or new investments than about fulfilling neoliberal reform plans. Thanks to a gradually improving global shipbuilding economy, both shipyards were once again receiving increasing numbers of orders. They were able to secure prestigious contracts from large international ferry and cruise carriers as well as orders for fishing boats and special vessels such as the live cattle transporter built by Uljanik, which could hold up to 75,000 sheep. The economic rollercoaster ride in Gdynia and Pula experienced another upswing, with both shipyards generating profits – Gdynia starting in 1997, and Pula beginning in 2000. During this period, Stocznia Gdynia became one of the ten largest ship manufacturers worldwide, and even considered acquiring two competitors in Finland. This business uptick meant more radical restructuring plans were laid to rest.

For a brief while, it seemed as though the two shipyards – and in fact, the entire shipbuilding industry in Poland and Croatia – might be able to return to their former (at least self-ascribed) glory. Janusz Szlanta, CEO of the Stocznia Gdynia since 1997, was elected "Manager of the Year" in 2000.[42] He had achieved the seemingly impossible by transforming a massively unprofitable public company into a profit-yielding private enterprise. The better-known shipyard in Gdańsk, on the other hand, was sliding inexorably towards bankruptcy or a takeover. Szlanta, the first director of Gdynia not to have a degree in engineering or shipbuilding – he was previously a financial manager – celebrated this turnaround in media interviews and during the annual press conferences presenting the company's results. The profits were relatively small, of course, with most of the proceeds being reinvested into the plant itself. But the shareholders did not expect big dividends, just an absence of losses. Szlanta personally would have been significantly affected by a negative balance sheet himself, since he had acquired a share of 20 per cent in the company in 1997 via an investment fund. He was the company's last universally recognized patriarch, and his tenure in Gdynia is considered the shipyard's "golden age" to this day.

A similar role was played in Pula by the engineer Karlo Radolović, who served as Uljanik's director until 2006 after having worked there since 1966. He enjoyed the loyalty of the staff but also knew his way around the political stage.[43] However, the large volume of incoming contracts belied an increasing number of problems. Full order books did not necessarily translate into profits. The orders for specialized ships often required the construction of prototypes, and it was difficult to recoup the immense development costs. The Polish and Croatian shipyards still suffered from excessive staff numbers and correspondingly low productivity compared to their Western and East Asian competitors. This caused Stocznia Gdynia to slide into the red again in 2003.

In Pula, too, the global economic and financial crisis of 2008/9 caused the erstwhile boom to be followed by the proverbial bust.

But did the two shipyards' eventful histories really end with them perishing in the stormy waters of transformation in 2009 and 2018, respectively? In Pula, shipbuilding – and thus one of Istria's last strongholds of industry – seems to be facing the final curtain. The only serious prospective buyer for the former Uljanik shipyard is currently an international tourism group that hopes to build a gigantic hotel resort and marina for luxury yachts on the site. Ultimately, it will come down to a choice between tourism and industrial production, and all the signs in this beautiful coastal area currently point to a triumph of the former. The term *deindustrialization*, used in the literature for this type of development, fails to capture the entirety of the associated processes, however. Besides the shipyard itself, the local shipbuilding environment comprises a host of educational facilities, including a specialized technical school in Pula that, in turn, maintains close ties with the Faculty of Mechanical Engineering and Shipbuilding at the University of Zagreb. If the shipyard were to disappear, the foundation and purpose of these institutions would be obsolete, and the notion of "structural change" seems insufficient for such a profound Polanyian transformation.

In Gdynia, on the other hand, the end of the large shipyard actually marked a new beginning for shipbuilding – one that in fact can be traced back to the end of state socialism itself. In 1990, two engineers founded the private shipbuilding enterprise Crist S.A., which limped along for a considerable time, mostly living off subcontracted work from the much larger public enterprise. In 2009, however, the company jumped on an opportunity afforded to it in a scenario of "creative destruction," a term coined by Joseph Schumpeter. Crist acquired the majority of the production facilities from the bankruptcy assets of Gdynia Shipyard, including the dry dock. It had generated enough capital for this venture through repairs, the manufacture of components such as bridges, specialized jobs including ocean floor installations, and its own conversion into a stock corporation. These niche sectors offered better profits than the general shipbuilding market. Naturally, besides the production facilities themselves, the rise of Crist – which does not consider itself a successor of Stocznia Gdynia[44] – relied on know-how, skills, and knowledge inherited from the state socialist period and the existence of a local shipbuilding milieu. The engineers and specialized workers the company needed were available in large numbers thanks to the gradual reduction of the workforce and eventual bankruptcy of its larger competitor. Socialist shipbuilding, thus, left not only a physical legacy that can be used for new purposes, but also a reservoir of skilled workers.

Social Advancement and Differentiation

Under state socialism the major enterprises were far more than pure production facilities. They were at the centre of their employees' lives, often across multiple generations. In our biographical interviews, we met many workers who had followed in their parents' and grandparents' footsteps. Nevertheless, such family traditions cannot conceal the differences between the generations that were already visible during the era of state socialism. For the first generation that began working during the Stalinist and the Yugoslav industrialization drive of the 1950s, shipbuilding meant social advancement. As in Western Europe, the economic and social *trentes glorieuses* of heavy industry lasted well into the 1970s, at which point the socialist regimes increasingly began to promote mass consumption and initiated a process of modernization financed with Western loans.

Many of these first-generation labourers after the Second World War came from rural areas of Istria and Kashubia – in other words, the hinterlands of Pula and Gdynia. They were fleeing widespread poverty in the villages resulting from high birth rates and the dire state of farming. In contrast to other industries, the shipyard workers received decent wages, apartments (after something of a wait), paid leave, and Sundays off – a luxury not afforded to agricultural smallholders. For the female workers, the shipyards often provided the first paid occupation they had ever had.

The children of this generation, who represented the oldest individuals in our group of interlocutors, usually joined the companies of their parents as a matter of course. For them, this no longer equated to social advancement, however, and a significant number therefore aspired to exchange the blue collar of the hard-working labourer for the white collar of the administrative employee. Nevertheless, jobs at Uljanik and the "Paris Commune" Shipyard remained in high demand, not least because the shipyards had become providers of a wide range of comforts and amenities. What was more, there were several specialized technical schools and colleges that allowed many male workers – and some of their female colleagues – to attain additional qualifications and thus further social advancement. Although it has been largely forgotten in the West, major companies such as General Motors in the United States or Volkswagen in Germany fulfilled similar all-inclusive functions in the first decades after 1945.

Our research was particularly interested in the developments beginning in the 1980s, when the third postwar generation began working in the shipyards – or by this time more often elsewhere. For example, in

Pula it became more and more difficult to kindle enthusiasm in students for an education at the naval engineering school followed by work at Uljanik. In Gdynia, the number of staff at the Paris Commune Shipyard fell by around 5,000 workers from its maximum of 12,700 in 1977 till the end of the 1980s. The shrinking of the workforce shows that the goals of efficiency and productivity were already being pursued well before capitalism officially arrived. Uljanik refrained from such staff cuts, thereby providing further proof of the differences between state socialism and self-management socialism, which was considered the great ideal among proponents of the Solidarność movement in 1980/1. While the democratic credentials of the Yugoslav self-management model are debatable, it did improve workers' bargaining power and increased their emotional attachment to the enterprise.

The shipyards remained attractive employers in the 1990s, if only because any work was better than unemployment, which was rising rapidly in both Poland and Croatia. Though they were no longer the showcase socialist enterprises they had been and also gradually lost many of their welfare functions, they were still the largest employers in the region. Initially, women's jobs were particularly under threat in the shipyards because they predominantly worked in administration and those branches then considered to be outside the "core business." However, once the economies of Poland and Croatia stabilized and alternative employment opportunities appeared, also abroad, the situation changed for all the shipyard workers. In view of the increasing uncertainty regarding the future of the shipyards, the third and fourth post-war generations pursued new career paths. Various other socio-economic factors, including significantly decreasing birth rates, the accessibility of new educational options, freedom of movement within the EU, and the ever-widening wage gap between blue-collar and white-collar employees, also contributed to the end of the worker dynasties.

Falling staff numbers meant the shipyards lost their former relevance for the local and regional job markets – and also their symbolic importance. Uljanik and Stocznia Gdynia had symbolized the promise of progress and a modern industrial society under state socialism, which the post-communist countries attempted to carry forward in a different way. For most of the shipyard staff, progress was not an abstract term; it was a part of their life experience. Around the turn of the millennium, when both companies were flourishing and the transformation to the market economy seemed to have been successfully completed, there was tangible progress at the material level as well. Real wages grew, as did consumption and purchasing power – not least because the workers' families were still living inexpensively in the now privatized former company apartments.

Did this growing prosperity make people happy, or at least more satisfied? More importantly, did it result in an emotional identification with democratic values and the neoliberal market system? There was a lively debate about these questions in Poland after the political shift towards the right-wing populist and nationalist party Prawo i Sprawiedliwość (Law and Justice, PiS) in the 2015 elections.[45] The polemics against consensus-based decision-making and the previous reform policies made it more difficult to obtain information on the individual and collective experiences from the time before the change of government during our interviews. Many of the answers we received were influenced or predetermined by the political polarization practised by the PiS. Notwithstanding this methodological problem, it seems clear that many ordinary people had mixed feelings about the successes and failures of transformation. In as early as 2009, a memoir competition for young people held by the KARTA Center in Poland on the occasion of the twentieth anniversary of 1989 revealed divergent opinions on the transformation – even though neoliberal economic reforms were still widely recognized as a success model in Poland at this point. The adolescents of the time largely saw material improvements during the 1990s and 2000s, for example in terms of new household appliances or cars. But they also spoke of massive insecurity among their parents and families as a result of business closures, rationalization, unemployment, labour migration, and the associated social problems.[46]

Pula and Gdynia alike were only mildly affected by the downsides of the transformation, at least in comparison to the marginalized parts of Croatia and Poland, which were generally in the east of the country. But the focus of production progressively changed, as lamented by the workers, specialists, and engineers interviewed by our authors Stefan Petrungaro, Andrew Hodges, and Piotr Filipkowski. For example, our interviewees complained that they were now reduced to designing and manufacturing ship parts as opposed to entire vessels. This is a further indication of the fact that the construction of ships in the era of high modernity held a significance beyond pure economic value – it was also about producing meaning. The all-encompassing, integrated organization of production under socialism, in which enterprises aimed to make themselves as independent of (unreliable) suppliers as possible, was reflected in the workers' holistic identification with "their" company. The loss of this holism between production and the meaning of life is something our interlocutors consistently bemoaned.

We have chosen to emphasize this finding with regard to symbolic production not least because it is relevant for all regions of Europe – including those in the West – that were subject to structural change. Workers' nostalgic memories of the period of state socialism are related not only to a perceived social relegation but also to a loss of identification with their entire company and its products – in this case ships, the largest machines created by man. The questions we asked our interviewees regarding their experiences since the mid-1990s were often answered in terms of feelings of estrangement: life at the company had felt different, and co-workers had subsequently begun to behave differently as well. The changes also caused a loss of trust in society because the organization of labour was stripped of its communal features, with the ritual of the ship's christening as one of its most important manifestations.[47]

These vernacular evaluations often rest on a strong binary between us and them. Now, it is clear that the working class was never a homogenous group, even though "labour history" originally implied a certain uniformity of experience and social status. Communist regimes had a keen political interest in promoting such a homogenous vision. Regardless of actual differences and inequalities among the imagined working class, the communist policies helped to generate class loyalties and we are all familiar with how this turned out in Poland: the protests, strikes, and revolts of 1956, 1970, 1976, and 1980/1 were organized and supported primarily by workers. Gdynia remains a central place of remembrance in Polish national history to this day; it was here that the army opened fire on a group of strikers on 17 December 1970, killing eighteen and wounding more than one thousand (along with forty-five further victims in Szczecin, Gdańsk, and Elbląg). This day is commemorated as Black Thursday (see figure 1.3).

This massacre, along with frequent other acts of violence by security forces, reinforced the proverbial separation of "us" and "them" (*my i oni*) in Poland, thereby inadvertently strengthening the sense of community. The "*oni*" were the representatives of the regime, although this primarily political delimitation also included a distinction between "up there" and "down here." This situation is not immediately apparent in the available written documents, however, as there were party members among the labourers as well as Solidarność supporters in senior management. But the community spirit among the industrial workers was further strengthened in 1980/1 thanks to their joint experience of political mobilization – and it would remain an important resource after 1989. One expression of this camaraderie was the fact that even specialists, engineers, and white-collar employees actively identified as *robotnicy* (workers).

Figure 1.3. March by Gdynia Shipyard workers to commemorate "Black Thursday," 17 December 1989
Source: Courtesy of European Solidarity Centre, © Wojciech Milewski.

In Yugoslavia, a tradition of corporatism and the principle of self-management embedded socialism more deeply in society. For a long time, the imagined antagonism between those "up there" and those "down here" – in other words, the party supporters and members of the opposition – were less pronounced than in Poland. After the 1950s, the League of Communists did not hold much significance in the workplace and was thus less visible in the everyday life of the workers. Self-management had conflicting consequences with regard to the workers' identification with their place of work, however, since it promoted a form of local self-will or even egoism. While the workers' councils failed to bring about true industrial democracy, they produced numerous stakeholders with an interest in the well-being of individual companies and provided a basis for local alliances between workers and directors.[48] As a result, decisions were more localized than in Poland – but so were protests: in Yugoslavia, there were thousands of strikes over the decades, with at least one being held at Uljanik in 1967.[49] Unlike in Poland, the regime did not interfere directly and such labour disputes were generally resolved quickly by way of local, company-specific concessions. As our interviews conducted in Pula show, the workers had a strong sense of self-confidence, owed in part to the idealization of the *radnici* (workers) by the government and the prevailing ideology. An overarching mobilizing force like the one in Poland was lacking, however. The Yugoslav regime consequently experienced no marshalling of workers comparable to what the Solidarność movement had achieved.

At both locations, which once again serve as allegories for their respective countries in this regard, the idealization of the working class was embodied in attitudes, actions, and a certain habitus, and generally foregrounded in iconography and public perception during communist rule. However, this social prestige of workers soon vanished after 1989. In Poland, workers in large factories were no longer seen as opposition fighters, but instead increasingly as relics of the past and freeloaders resisting the necessary reforms. In Croatia, too, mainstream political discourse began to view industrial workers as more of a burden than a force of progress. This stigmatization was partly an intentional strategy pursued by neoliberal reformers in order to generate public approval for radical restructuring plans and job cuts. At the local level, however, many of the amenities and practices that had shaped the workers' identity and sociability – including their spatial concentration in neighbourhoods with many still living in (former) company apartments and meeting for drinks in the ubiquitous, nondescript cafés – persisted regardless. The resulting consequences for the attitudes and sense of community of the workers will be examined by means of interviews, ethnographic observations, and document analysis.

Our team was particularly interested in the growing differentiation – or perhaps fragmentation – of the working class during the period of the "long transformation" and how this developed in connection with political and economic changes. For example, internal differences emerged as a result of the influence of divergent professional qualifications on individual fates when job cuts were carried out. Shipbuilding is characterized by a strong tradition of trades-related identities. This means that specialized workers such as welders and painters did not necessarily consider themselves members of the same class. From the mid-1990s, the different status of workers at the shipyards was increasingly evident in the growing wage gap and different consumption behaviours, with the make and size of the car a worker owned becoming a particularly clear symbol of the inequalities.

Uncertainty resulting from the economic decline of the shipyards, which became increasingly obvious in Poland starting in 2003 and in Croatia from 2009, gave rise to an even more pronounced social and cultural differentiation among the members of the youngest generation. As mentioned earlier, the children of shipyard employees increasingly followed other career paths: they went to university instead of learning a trade at the shipyard, moved to other cities or abroad, established their own companies, or took up jobs in tourism or the service sector. In the worst case, they wound up in precarious living conditions. Put differently, one of the reasons why the life plans and biographies of the people of Pula and Gdynia are becoming more diverse is that even prior to their economic failure, the shipyards' defining symbolic power and their dominance over people's lives had begun to dwindle – both inside and outside the factory gates.

In the past, these enterprises had offered acceptable employment opportunities along with access to holiday homes, inexpensive housing loans or company apartments, and the chance to engage in various sports or support the shipyard's team as it vied for success in a sports league. The workers could also spend their leisure time in the cultural institutions maintained by the company – Uljanik even boasted its very own rock club. The transformation of the shipyards in the 1990s soon left only the job itself, and then often not even that. Accordingly, the significance of shipbuilding and work at the shipyard within the emotional world of the populations of Pula and Gdynia changed, as did the role of the industry in the local geography of work. Our combination of sociological and historical approaches enabled us not only to observe these transformations in real time, but also to historicize them and investigate their continuities and path dependencies.

The bankruptcy assets of the former shipbuilding companies remain important economic and cultural resources in Pula and Gdynia to this day. They define the image of the cities as well as their collective memories – and in Gdynia, they represent the economic basis for Crist as well. In this sense, the history of naval architecture in Gdynia did not come to an end in 2009. On the contrary, considerable hope for the future rests on new developments such as the first fully electric ferry boat, built and presented in Gdynia in 2017.[50] Whether a similar continuation of the long history of shipbuilding in Pula, which can be traced back to 1856, will be possible remains to be seen. The conditions for a revival are less favourable there due to the lack of an industrial environment like the one enjoyed by Gdynia with its proximity to Gdańsk, its various links to Germany and across the Baltic Sea to the Nordic countries, and its own facilities nearby.

The most likely scenario in Pula at the moment is the acquisition of the shipyard by a hotel group, though it is likely that this too will take many years owing to complex ownership issues. What is more, nobody knows what might be found in and underneath a plot used intensively by an industrial facility for more than 160 years, and for several decades by the military. There are numerous indications that Pula will follow the path taken by many Mediterranean towns and become a tourist "monocity." However, the tourism industry is restricted to specific seasons and largely dependent on the purchasing power of people living elsewhere – namely, in the economic centres of Europe, over which Croatia has no influence. The experience of the Covid-19 pandemic has demonstrated how vulnerable this sector is.

Can robust conclusions for the future of Europe as a location of industrial production in general or for the further economic evolution of the post-socialist countries be drawn from these experiences in the shipbuilding sector? If so, then developments since the late 1970s offer little room for optimism. Europe as a whole has slowly been drifting towards the periphery of the global economy, at least in manufacturing sectors such as shipbuilding.

Methodology and Sources

Can such general conclusions be drawn from two inevitably very specific case studies? We discussed the problem of generalization and validity in our project from the very outset; this is something that has also been the subject of controversy ever since the establishment of oral history and microhistory. Of course, we cannot generalize too much from two single enterprises and their labour relations. Nevertheless, our cases do reveal a lot about the long afterlife of socialist modernity, the impact of economic doctrines and policies after the end of communism on everyday life, ruptures in social structures, and the shift in labour conditions and shop floor relations.

In addition, the shipyards illustrate the impact of Poland's and Croatia's EU accession on these countries' economic transformation. Using these businesses as examples, we can show that this specific turning point had more fundamental consequences than existing literature on the contemporary history of the two countries acknowledges. Our in-depth analysis revealed further surprising aspects of a more general nature, such as a shift from state socialism to state capitalism rather than the common paradigm of privatization. The latter nevertheless remained compulsory as a speech act. This would change only after the great crisis of 2008/9, when the neoliberal rhetoric lost some of its credibility, while at the same time the EU forced the two countries to finally privatize the last of the factories that remained in state hands.

We also repeatedly asked ourselves whether around fifty biographical interviews, along with expert interviews, allow inferences concerning the respective enterprises and employment relations. We attempted to address this methodological and theoretical concern by speaking to a wide range of people, from the shop floor all the way up to former managers and important state actors, and by encompassing as diverse a sample as possible in terms of generations, genders, and occupational and social positions. The impressions, feelings, and experiences shared by our interview partners are naturally always individual – but the repetitions, contradictions, and omissions observed in the totality of the conversations add up to a collective narrative that fulfils the requirement of describing the transformation "from below," meaning from a biographical perspective as stated in our project title.

In terms of the number of interviewees, we spoke with roughly as many individuals as various other oral history studies that are now considered classics.[51] The criticism occasionally levelled against oral history or qualitative sociology – namely, that biographical interviews are too subjective – can be rebutted with confidence: more voices and pieces of information are included in our portrayal than in seemingly objective corporate data, which we obviously also took into consideration. Besides, the focus on different subjective positions and experiences was necessary to introduce nuance to group categories such as "workers" that we also employed ourselves. In Poland and Croatia, only the male form – *robotnik* and *radnik* – was used, stressing the masculinity of the shipyard workers and marginalizing their female colleagues.

A further theoretical problem is the considerable amount of time that has lapsed since the historical processes our team of sociologists, anthropologists, and contemporary historians investigated.[52] Until the 1980s, proponents of oral history assumed that interviews were appropriate tools for tapping into authentic experiences and writing a genuine history "from below."[53] But the linguistic turn increased awareness of how problematic the idea of reconstructing historic "realities" based solely on interviews actually is.[54] Ultimately, such work always deals with narratives predetermined by the political, social, and cultural context at the time of questioning, by the life experiences of the interviewees, as well as by the specific questions asked. As a result, the researcher may end up learning more about the *view* of the past in question than about actual historical events and processes – the totality of which may not even be fully representable at all. But it is precisely the fact that these interviews reflect the multifaceted experiences of people in Poland and Croatia during the past decades and the diverse circumstances of their current lives that makes them such valuable sources for a project like ours: the "great" transformation entails many "small" transformations, and vice versa. In other words, the transformation represents more than just a macrosociological process; it is also a shift in frameworks of meaning – a space of experience that can be grasped only by reconstructing the perspectives of individuals and small communities. Aside from this, we are also convinced that despite the narrative quality of biographies, interviews still constitute valuable references to past events and help to close certain gaps that printed or archival documentation leaves open.

Initially, our research as well as this book were focused on a comparison in terms of business history, along with the question of governance from above and the state as a business operator. In this way, we explicitly hoped to confront the holistic and largely sceptical view of the role of state that has increasingly prevailed since the 1980s as a result of the global hegemony of neoliberalism. Milton Friedman and the Eastern European adepts of the Washington Consensus demanded that "the state" withdraw from the economy and concentrate on its narrowly defined primary tasks (similar to the way companies were urged to focus on their "core business"). Although we likewise use "the state" as an abstract term, we do not view it as an abstract unit or singular agent but instead try to distinguish different levels and agents of statehood. Based on this approach, the silhouette of the "visible hand" of the state emerges quite clearly – and neoliberal reforms naturally depended on government intervention and regulation as well. Most importantly, we considered a level of statehood that has previously barely featured in comparable studies, despite being central to the fundamental changes experienced by East and West alike over the past decades: the European Union.[55]

We attempted to address the aforementioned theoretical challenges with a broad mixture of sources as well as of historical and sociological research methods. Consequently, we followed a paradigm inspired by historical anthropology and historical sociology that aims to understand the structures within which humans act and to interpret the frameworks of meaning and spaces of action that – sometimes consciously, but more often unconsciously – affect and alter these structures. Anthony Gidden's idea of "structuration" succinctly describes our approach: structures cannot exist without agency, and vice versa.[56]

In addition to our biographical interviews, we used internal transcripts and documents from the shipyards and a variety of government bodies as well as media sources.[57] The latter included topical articles on the shipyards from regional and national newspapers, recent interviews with their management staff and representatives of the state, industry publications and expert reports from the period, and of course Uljanik's and Stocznia Gdynia's extensive and informative company magazines. While the contents of these self-published periodicals were more heavily monitored, they were also more detailed and more focused on the experiences and attitudes of the staff than articles in daily newspapers, which generally dealt only with the high and low points of the companies' histories as well as results and figures (for an example, see figure 1.4).

Of course, the balance sheet numbers were of interest to us from a business history perspective as well, but we did not assume these profit and loss statements to present objective data. Socialist bookkeeping worked according to its own rules, and even during the era of capitalism, the books were visibly "cooked." And even when this was not the case, the system of bookkeeping was based on specific rules and frames governing how to represent the present and project the future of a business. A key question for us in this context was at what point profitability began to predominate over other success indicators and interests, and how success was measured.

The sheer volume of documents relating to shipbuilding in general and our two shipyards in particular is indicative of their significance, at least in the eyes of government institutions. It also showcases the close ties between the shipyards and the state: many of the official papers and letters concern the companies' repeated requests for government aid. We were also granted access to the corporate archives of the two shipyards. In Poland, this was possible owing to the fact that Stocznia Gdynia did not destroy the majority of its archived material following its insolvency, instead donating it to the regional state archive in Gdańsk (Gdynia branch). Uljanik in Croatia was similarly generous, even though the company was still operating at the time of our research. We were granted access to the vast archive located in the basement of the shipyard's former administrative building. This culture of openness is one of the positive legacies of 1989. In Germany or indeed any other Western country, such recent materials, including important economic and personnel data, would certainly have been off limits, especially in company archives.

In combination with our sociological interviews, access to very recent and sometimes even current archive holdings, together with our own extensive archive of media coverage, provided unique insights. The "triangulation" of information and perspectives allowed us to reconstruct specific events and circumstances using different written and oral sources. In doing so, we noticed that sources generally considered to be

Figure 1.4. Cover page of Uljanik company newspaper, January 1986
Source: Courtesy of Historical and Maritime Museum of Istria.

subjective sometimes capture historical processes more accurately than putatively objective documents. A further distinctive methodological trait of our project is owed to the research traditions of Polish sociology. Piotr Filipkowski was able to find respondents to large-scale sociological surveys reaching back as far as the 1970s. While the results of these surveys had originally remained unpublished due to censorship and content that was potentially critical of the regime, our research team was allowed to access these formerly confidential documents. What is more, the surveys were resumed in the 1990s and only recently discontinued. This enabled us to reconstruct the ways in which attitudes regarding the political and economic system – as well as individual experiences in people's professional and private lives – have changed over time. In addition, sociologist Peter Wegenschimmel conducted interviews with former CEOs and ministers, thereby obtaining insights into the highest echelons of leadership. Based on these sources, we were able to examine memories of state socialism and the upheavals of 1989 and 1991 not only from the current perspective but over the course of time – and thus in a more profoundly "historical" sense.

The fact that we owe this opportunity to a historical sociologist is testament to the value of interdisciplinary work. After all, our project's fundamental goal was to historicize the transformation without losing sight of its tremendous sociological relevance. In Pula, Stefano Petrungaro and Ulf Brunnbauer, social historians like Philipp Ther, also conducted numerous interviews that would be more accurately classified as sociological rather than oral history sources owing to the frequently contemporary nature of their content. The skills of cultural and social anthropologist Andrew Hodges, whose work alternated between interviews and ethnographic field research, were similarly beneficial. Perhaps inspired by the desire for community so apparent in the two cities studied within our project, our intent was to generate a "multigraph" of sorts, created by a collective of authors from various disciplines and exceeding – in terms of breadth and quality – what any individual researcher could have achieved with a monograph.

Finally, a few words on the heuristics of the comparison are in order. Our study presents a binary and contrasting comparison with the primary purpose of carving out the differences between the two shipyards studied and their respective national contexts.[58] However, a number of commonalities became apparent as well. These partly resulted from the globalized nature of the shipbuilding industry and its technological prerequisites, and partly from the development of the two companies themselves. To avoid the static nature of many comparisons, we have attempted to chart these differences and commonalities as divergences and convergences. In very general terms, we detected an increasing convergence between Uljanik and Stocznia Gdynia. During the socialist period, the two shipyards were organized and positioned along very different lines, and relations among their respective staff were quite distinct as well. The reasons for these differences were path dependencies in the two shipyards' company histories, as well as – and more importantly – the type of economic organization implemented by the Polish and Yugoslav communists. The normative frameworks in Yugoslavia and Poland, within which the two shipyards operated, were quite different. Uljanik and the entire shipbuilding industry along the Adriatic coast initially had a considerable head start in terms of integration in the world market – and this advantage once again stood *pars pro toto* for Yugoslavia. Poland, on the other hand, was then clearly situated on the periphery of the European and global economy, being heavily dependent on the Soviet Union.

The collapse of Yugoslavia and the subsequent war – both important external factors for the Uljanik shipyard – nullified this advantage. A convergence between Uljanik and Stocznia Gdynia begins to emerge in the mid-1990s owing to the context of actually existing capitalism, on the one hand, and the EU and its common regulatory frameworks, on the other. It ultimately also manifests in the shipyards' common fate of insolvency – which occurred more or less exactly five years after each country's accession to the EU. Comparative studies sharpen the focus on the peculiarities of the individual study objects and their contexts that might otherwise have been overlooked; commonalities allow us to draw generalizing conclusions that extend beyond the two entities being studied.

Traditional comparisons have often suffered from the assumption that the objects of comparison exist in isolation from each other. In times of entangled history and *histoire croisée*, however, it is important to maintain an awareness of boundary-transcending links between entities. Key members of Solidarność, for example, considered Yugoslav self-managed socialism a role model in 1980/1. Yugoslav and Croatian shipbuilding experts, in turn, followed the developments in the sector in Poland with great interest. As mentioned earlier, this close relationship was also a result of the heavily globalized shipbuilding market. Although the two shipyards and shipbuilding nations competed with one another in certain market segments, they did not really perceive each other as rivals. This was not least a result of the shipyards' conscious efforts to develop their own niches. Seen from a global perspective, Uljanik and Stocznia Gdynia were in the same proverbial boat as other European countries – being increasingly outpaced by the East Asian (and especially Chinese) competition over the past fifty years. In this sense, the two shipyards are ultimately emblematic of changes in global industrial geography and Europe's weakened position within it.

Looking at the EU as a whole, this weakness is linked to an intellectual and political vacuum, to the inability and unwillingness to see the state – or at least specific state institutions – as entrepreneurial entities. The governments of the nation states of Poland and Croatia repeatedly extended a helping hand, but they never truly controlled the companies or stimulated them to perform. These issues are not new, of course; the state-run industries and industrial holdings in Western Europe were in a similar situation in the 1980s. In this context, the vacuum was filled by way of a twofold approach: on the one hand, more or less successful privatizations in keeping with the economic schools of thought in Chicago, Freiburg, Milano, and so on that dominated at the time, and the shifting of shipbuilding and other industries to East Asia, on the other. Here, the state not only extended its helping hand but actively pursued industrial and development policies that entailed far-reaching consequences for the European production sites – first in the West, then in the East. Should the EU, which is not a state per se but has successively accumulated more and more state-like authority, restrict itself to the role of an inner-European competition regulator? In light of the unprecedented economic challenges caused by the Covid-19 pandemic, the Russian war against Ukraine, and the necessary "rebuilding" of Europe – with all the consequences this has for the future of labour, reaching into the finest capillaries of society itself – this question appears more relevant than ever.

2 Forever on the Verge of Going Under: A Tale of Two Shipyards

On 28 August 1978, a delegation headed by then Chairman of the Chinese Communist Party (CCP), Hua Guofeng, visited the Uljanik shipyard in Pula (for the outline of the shipyard, see map 2.1). Guofeng's official visit and his meeting with Tito served not only to strengthen the friendship between Yugoslavia and the People's Republic of China (PRC); there were also more tangible matters to be discussed. Indeed, during the visit, the PRC placed its first order with Uljanik for five ships and twenty marine engines. The shipyard's director, Karlo Bilić, was of course very happy with this outcome, expressing his hope "that that cooperation with the People's Republic of China would continue to grow and thrive in their mutual interests." Guofeng diplomatically extolled the virtues of Uljanik's 120-year history and Yugoslavia's "highly developed shipbuilding industry."[1] Reporting on the visit, the local newspaper *Glas Istre* emphasized that business from China could see Uljanik working at full capacity until the end of 1980.[2] This was welcome news indeed, given that the shipyard was struggling to find customers in the capitalist world owing to the lasting decline in demand for new ships since the oil price shock of 1974. At the time of Guofeng's visit, the shipbuilding industry in the PRC was rather rudimentary.[3] However, the Chinese leadership had plans to expand the sector, seeing it as an incubator of industrialization that could fuel progress in other branches of industry and promote foreign trade.

Around forty years later, in early May 2019, Pula welcomed another Chinese delegation. Except this time, it was not members of the CCP who paid the city a visit but representatives of China's biggest shipping company, Cosco. These Chinese delegates did not, however, come to place orders with the shipyard or to learn from it. Instead – at least this is what Uljanik and the local government were pinning their hopes on – they were its potential saviours. At the time of this visit, the shipyard

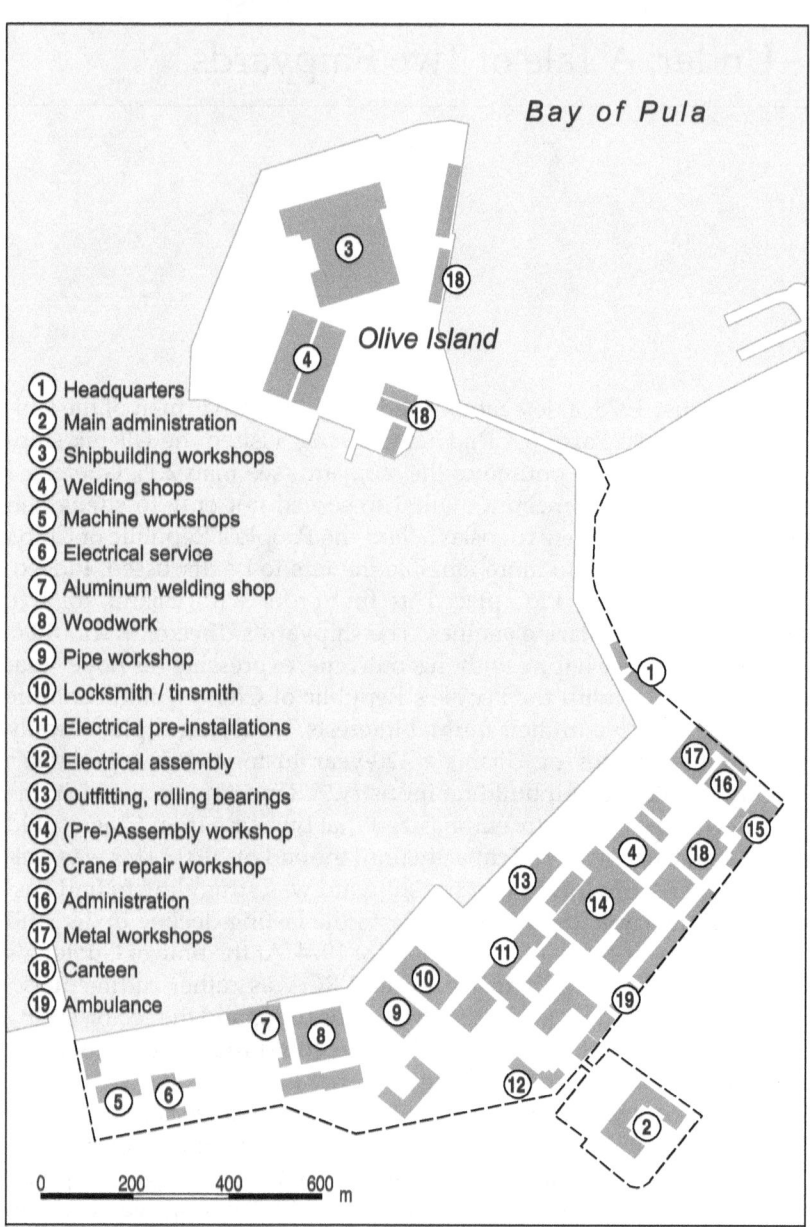

Map 2.1. Outline map of Uljanik shipyard, 1990

had been in its death throes for a year and a half, with production halted due to lack of funds and an insolvency application already filed with the courts. Neither domestic nor European investors were willing to take over the loss-making shipyard. Customers had cancelled their orders and employees – at least those who had not already found employment elsewhere – were condemned to inactivity on a minimum wage provided by an emergency fund set up by the government.

So why all of a sudden were Chinese investors making an appearance? One month earlier, at the beginning of April, the fifth summit of the 16+1 Initiative had taken place in Dubrovnik. This transregional platform was launched by China with a view to stepping up its cooperation with the countries of Central and Eastern Europe as part of its Belt and Road Initiative, also known as the "New Silk Road." On the sidelines of the summit, Croatian prime minister Andrej Plenković told his Chinese counterpart Li Keqiang all about the sorry state of Croatia's two biggest shipyards, Uljanik in Pula and 3. Maj in Rijeka (at the time the Rijeka shipyard also belonged to Uljanik). Plenković brought up the possible entry of Chinese investors and just a few weeks later, the aforementioned representatives of state-owned enterprise Cosco arrived in Pula.

Unlike in 1978, when the Chinese delegation had been highly impressed by the calibre of Yugoslav shipbuilding, the 2019 visit saw them become rapidly disillusioned. *Glas Istre* reported that the "Chinese felt as though they were in a shipbuilding museum."[4] They had never seen cranes as old as Uljanik's – their own shipyards and shipyard equipment being twenty years old at most. After a few hours of niceties, it became clear that the Chinese visitors had no interest whatsoever in investing in Uljanik, let alone in buying the two shipyards in Pula and Rijeka, also owned by Uljanik. A Chinese bailout would have been the final ironic twist in Uljanik's fateful tale, not least because the rise of China's shipbuilding had been one of the main factors contributing to the decline of the European shipbuilding industry. In 1978, when Chairman Hua Guofeng visited Uljanik, the PRC's share of the international market for commercial shipbuilding was exactly 0 per cent – by 2017 it had climbed to 35 per cent.[5] Given Cosco's huge capacity, there was not a single good reason for them to invest in the Croatian shipbuilding company – an enterprise that was, by Chinese standards, far too small and ailing. Although today's China very clearly recognizes that trade policy is tantamount to foreign policy, Chinese companies still want to make money out of their foreign activities. No more than a month later, the district court in Pazin opened insolvency proceedings against Uljanik.

This is not to say, however, that Uljanik is an exception in Eastern Europe. In fact, ten years earlier, Stocznia Gdynia (formerly the Paris Commune Shipyard) – at the time the biggest shipyard in Poland and our comparative example – was forced to close its gates (the layout of the shipyard is shown in map 2.2). Both companies had been struggling for years, ultimately in vain, to adapt to the conditions of the capitalist market economy, wrestling with the competition from East Asia. The ten-year gap between the bankruptcies of the two companies obscures an important similarity: both shipyards went bankrupt five years after their home countries joined the EU (Poland in 2004, Croatia in 2013). This was not a temporal coincidence but was, as we will show, directly related to EU regulations. The current chapter, like the rest of the book, will therefore focus on the relationship between the two developments – EU accession and shipyard bankruptcy. The European Union's competition rules strictly limit the value of subsidies member state governments can give companies, in doing so restricting their capacity to protect domestic companies from global competition.

For Poland and Croatia, EU accession represented a whole new chapter – not only politically, but also in terms of the relationship between the state and companies as well as business practices in the broader sense. Their membership of the EU saw an end to a long process of transformation that had begun in the early 1970s. The two countries had entered a new era where there was clearly no longer any room for traditional shipyards and the modes of production and industrial relations they entailed. Here, we argue that shipbuilding does not, however, stand in isolation. In fact, the industry offers a heuristic added value for the entire process of transformation. The shipyards serve as a pivot that connects the different levels determining the path of transformation, including the global market, European policy, national and local developments, and technological change. Given the interaction between the protagonists on the different levels, all of these aspects influence the production of ships. The exact outcomes of this interaction depend on the location-specific constellations of protagonists as well as their institutional conditions. To allow for even more broadly applicable conclusions, we compare two different contexts, the Polish and the Yugoslav/Croatian, each with their own distinctive dynamics. Both cases saw a major shipyard go bankrupt, albeit each with a different form of "afterlife."

For some observers, it is not the demise of the two shipyards that warrants explanation but rather the duration of their survival. After all, these companies had made barely any profit since the late 1960s and repeatedly – if not permanently – found themselves in a state of

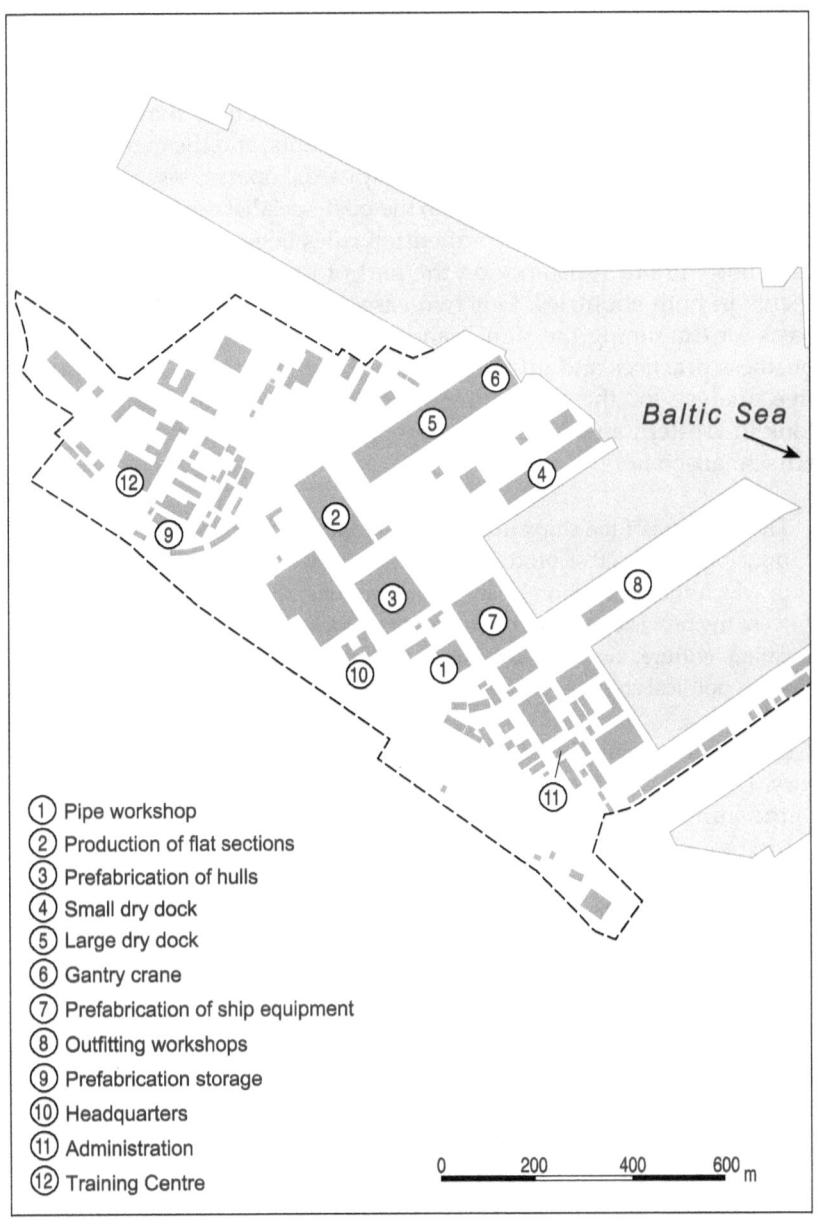

Map 2.2. Outline map of Gdynia Shipyard, 1995

financial distress.⁶ The answer to this puzzle lies in the specific added value that the two shipyards produced – and this was something that, for many years, did not in fact rest on the premise that the primary function of a company is to make profits from its production. Instead, it was social, but also symbolic and political objectives that were at the heart of what local actors, national governments, and the general public perceived as the actual value of the shipyards' operations.

This function prevailed well into the post-socialist era, indeed becoming obsolete only once EU competition rules began to apply, and even then not without resistance on the part of the public and the governments in both countries. Our two case studies thus provide an ideal basis for examining the significance of the end of communist rule for business practices and identifying possible continuities. In a comparative analysis of the different development paths that shipbuilding took in Western and Eastern Europe, Jürgen Bitzer and Christian von Hirschhausen note,

> The structure of the shipyards and the modes of production reflected the principles of socialist production: shipyards were multifunctional units, in which the production of ships was but one objective; other functions were the provision of social services to employees (such as housing, education, culture, access to consumer goods, etc.) and the maintenance of some political activity and control.⁷

For experts on planned economies, this observation is far from revelatory. The real, and far more important, question is how these functions of the shipyards could continue to influence – albeit to a lesser extent – business practices and the expectations of the factory community for so long after the radical change of the political system? Was this down to sheer inertia? Or was it more of a proactive attempt to preserve the social objectives, not allowing them to be swallowed up by the capitalist market economy in the pursuit of profits, while still managing to remain an actor on the global market?

In an attempt to answer this question, we will look at the web of relationships between the state and business, revealing astonishing continuities that challenge the widely held notions of the rapidly installed hegemony of post-1989 neoliberalism. We describe how the business of the two shipyards evolved in the context of the development of global demand for ships, the embedding of the companies in two different types of socialist economy, and Poland's and Croatia's processes of transformation. Our focus here will therefore be on the history of these companies and their key players and their interactions, which we conceptualize as a long process of "muddling through" – producing a transformation of sorts.

Transformation was not a linear, intentional transition from point A to point B. The transformation was in fact characterized by permanent processes of negotiation, by "driving by sight", by making (or failing to make) decisions in situations of overwhelming uncertainty, by tentative experimentation, or by reverting to what is familiar. The outcomes of these processes of transformational change were thus always unpredictable and did not match up with the actors' intentions. The transformation was marked by two specific events that can be understood as vertices, given that they both resulted in a genuinely radical change in structural conditions. The two developments we are referring to here are first, the shockwaves sent through the global economy by the two oil crises in the 1970s and second, the integration of Central Europe into the European Union.[8] At the time, many actors were not even fully aware of the transformative character of these historical events. In hindsight, however, their importance as turning points is as crystal clear as in the case of the 1989 revolutions. This statement certainly applies to shipbuilding but, more than that, it is our view that it also applies to (post-)socialist Europe in general. In other words, the transformation began long before the communist regimes fell and lasted much longer than the first reformers would have us believe.

Global Economic Cycles and the Late Socialist Transformation

In his pioneering book *Globalization under and after Socialism*, Besnik Pula argues that "the 1970s were a crucially transformative decade for socialist economies."[9] This applies all the more to the shipbuilding industry as, on the one hand, the demand for ships is closely connected to the development of global trade and, on the other, the market for merchant vessels is highly globalized, with producers in socialist countries competing with shipbuilders all over the world. The Uljanik and Gdynia shipyards both had to export because their production output far exceeded domestic demand. Indeed, their high export rate was one of the factors that drew particular political attention to the two companies.

The end of the Bretton Woods system in 1971, the oil price shocks in 1973/4 and 1979, the subsequent economic crisis and inflation, and the growing role of finance (partly as a result of the glut of petrodollars and the increasing importance of central banks for fiscal policy) brought about a structural change in the world economy. This was something the state socialist countries – and Yugoslavia in particular – proved unable to shield themselves from, precisely because the first half of the 1970s was marked by stronger international economic interdependence between the two blocs. During this period, Poland and Yugoslavia both took out substantial Western loans for the purpose of modernizing their ailing industry and increasing their exports to the West, as well as to help them subsidize domestic consumption and finance the socialist welfare state.[10]

Most of this new money flowed into the same old channels, that is, capital goods and heavy industry, areas the communist economic planners always saw as a priority. In contrast to Western Europe and the United States, where coal and steel – the former growth engines of their economies – lost importance, Poland and Yugoslavia continued to pursue a development strategy that largely depended on heavy industry.[11] In the West, on the other hand, the macroeconomic crisis of the 1970s accelerated the transition to more IT-based production, bringing huge gains in efficiency in its wake. While Western industrial enterprises were investing heavily in technologies that saved labour power and linked production steps more efficiently, the socialist countries were preserving an image of industry where success is measured by the number of people it employed and the number of smokestacks it had. The notion of investing in machinery and equipment that would see workers replaced and laid off was simply inconceivable for the communist regimes.[12] In the 1980s, the reluctance to reform and dependence on foreign donors took both Poland and Yugoslavia to the brink of bankruptcy – one of the key factors contributing to the collapse of communism.[13] With this in mind, Marie-Janine Calic sees the collapse of the Yugoslavian system as dating back to the 1970s – contrary to the popular belief that this was in fact a golden age. Calic argues that the political system was confronted with steadily increasing economic tensions and contradictions that it proved incapable of dealing with consistently, much less resolving.[14] This conclusion also applies to the Polish People's Republic, where the political system faced its final crisis at the very latest when martial law was imposed in the night from 12 to 13 December 1981. As Fritz Bartel noted, the communist regimes lacked the legitimacy to terminate the social contract (growing prosperity for the workers), while in the West – especially in Britain and the United States – the economic problems of the second half of the 1970s went on to bring about a breakthrough for neoliberal supply-side policies and austerity measures.[15]

Like under a magnifying glass, shipbuilding serves to highlight this dynamic, which was the product of macroeconomic developments and economic policy decisions. As social historian Marcel van der Linden notes, "Maritime trade is the backbone of the world's economy."[16] And given that it is ships that transport the goods, economic crises resulting from the decline of international trade will quickly impact the demand for

Figure 2.1. Ship engine workshop in Uljanik, mid-1950s
Source: Courtesy of Historical and Maritime Museum of Istria.

ships. This same dynamic also influenced our two case studies, which were by no means isolated from the global market. On the contrary, in fact, they both produced ships for export, Uljanik to an even greater extent than Gdynia Shipyard. For the latter, the Soviet Union was the most important customer, while, owing to Tito's active foreign policy, which always opened doors for Yugoslav firms, Uljanik exported to a number of Western countries and to Non-aligned Movement nations, as well as "flag of convenience" (FOC) countries such as Liberia and Panama. Thanks to generous state subsidies, both shipyards enjoyed a rapid comeback after the Second World War and were seen as excellent examples of modernization in what had to date been considered "backward" maritime regions.

Uljanik had profited from the liberalization of foreign trade (1952) as well as the export-oriented economic policy that Yugoslavia was already pursuing in the 1950s. The shipyard was able to procure licences from Western companies (for diesel engines, for example, see figure 2.1), completing, in 1958, the construction of its first ship for a foreign customer since 1945 – the new cargo vessel *Al Mokattam* for the Egyptians, who were close allies of Yugoslavia.[17] In the 1960s, thirty-six of the fifty vessels supplied by Uljanik were exported. Producing for export became the mantra of success ritualistically repeated by the management. By 1972, with revenues of almost a billion dinars, Uljanik was the country's leading industrial exporter, with no other shipbuilding nation in the world boasting as high an export rate as Yugoslavia.[18]

This was one way it differed from Poland, where half the vessels built at Gdynia Shipyard were destined for domestic customers. And when it came to Gdynia's export markets, the Soviet Union was the clear leader (with forty-four ships being built for customers from the USSR in this decade), although the company also had customers from India, Indonesia, China, France, Norway, Great Britain, and the United States on its books. In the context of the "socialist division of labour" within the Council for Mutual Economic Assistance (Comecon) established in 1949, Polish shipbuilding played a prominent role. For the Polish shipbuilders, the Soviet market, which was not in fact even a market, acted as a buffer against external shocks from the world market.

The global economic crisis, triggered by the two oil shocks, heralded the end of a long phase of expansion of shipbuilding around the globe that had begun in the 1950s with the rapid growth of international trade. The crisis hit shipbuilding particularly hard, as in the years preceding it, tanker construction had become the main business for many shipyards. Indeed, in the early 1970s, tanker vessels made up around 80 per cent of shipyards' orders worldwide, with both the number of ships and ship size dramatically increasing in the run up to the crisis.[19] In the wake of the 1973 oil price hike, demand for new vessels immediately plummeted, especially for tankers. In 1976, new order volumes worldwide had dropped to just 46 per cent of the 1974 level.[20] Prices fell by 40 per cent in a single year.[21] Scores of customers cancelled their orders, despite the contractual penalties, or negotiated new contractual conditions with the shipyards. Some simply refused to accept the vessels on completion or pay the final instalment. Then, in 1979, as the first signs of recovery just started to become visible, the second oil price shock plunged shipbuilding into a new crisis. Between 1976 and 1984, production in the leading Western European shipbuilding nations, which in international comparison were hit hardest by the crisis, fell by an average of 50 per cent, with Britain even recording a 70 per cent decline (South Korean production, in contrast, saw growth of 200 per cent, which is something we will come back to later).[22] It was not until the turn of the millennium that the industry reached the 1974 production level of around 36.4 million gross tonnes (overall internal volume of ships supplied). From 1978 to 1991, the global output for new vessels was less than half the volume recorded in 1974, a figure that did not begin to steadily grow again until 1990.[23]

In response to the crisis of the 1970s, governments around the world came to the aid of the shipbuilding industry, using various measures, including subsidies and import restrictions. In a memorandum published in 1977, the then government of the Socialist Republic of Croatia estimated that Western European governments were financing between 20 and 30 per cent of the price of a ship – in Britain, subsidies even made up as much as 50 per cent. This enabled shipyards to sell their ships on the world market at dumping prices and thus sustain their production.[24] Many Western producers profited either from direct protectionist measures or from state support for domestic shipping companies, making it possible for them to sell their ships to customers in their own countries.[25] The West German government, for example, gave its shipping companies a helping hand through its own shipbuilding assistance program (*Reederhilfe*) that, despite being in place since 1962, did not take on any real significance until this time and was in fact not discontinued until 1987. Through this program, in 1978, for example, German shipping companies received more than 400 million German marks in government subsidies "for the construction and upgrading of merchant ships."[26] In the late 1970s, the Swedish and British governments went as far as to nationalize the entire shipbuilding industry – a step that still failed to halt the demise of shipbuilding in either country.

In the West, government aid was often tied to certain conditions. The shipyards had to increase efficiency and reduce capacity. In a process that lasted fifteen years, around two-thirds of Europe's shipyards perished and with them 80 per cent of jobs in the sector.[27] In Britain, for instance, the number of people employed in shipbuilding fell from about 55,000 in 1975 to less than 15,000 just nine years later.[28] And in the West German shipbuilding industry, the number of workers in 1990 was a mere 32 per cent of the 1975 level.[29] Another adjustment process that was characteristic of the West German shipyards was the switch to the production of cruise ships, vessels for which they could charge relatively high prices, especially because Asian shipbuilders had not yet penetrated this market (thanks to the cruise ship tourism boom, this plan came to fruition, at least until the outbreak of the Covid-19 pandemic). Even Japan, the world's leading shipbuilding nation, which accounted for almost half the global output in the 1970s, reduced its capacity by one-third, investing instead in increasing efficiency. In 1975, a total of 256,000 people were employed in Japanese shipbuilding – just fifteen years later, the workforce had fallen to 89,000.[30]

The crisis of the 1970s undoubtedly accelerated the decline in importance of (Western) European shipbuilding compared to Asia. In the early 1950s, European shipyards built more than 80 per cent of the merchant vessels produced worldwide, but by 1990, this share had fallen to less than 20 per cent (further slumping to single digits after the turn of the millennium).[31] Perhaps no other country illustrates this development as clearly as the erstwhile global market leader, Great Britain. While in the 1930s, British shipyards still produced half of the tonnage built worldwide, in 1980 they were responsible for just 3 per cent, and by 1990, this figure had plummeted to less than 1 per cent.[32]

A crucial factor in this shift was the rise of South Korea, a country that, with the help of state subsidies and targeted industrial policy, managed to establish huge shipyards in the 1970s.[33] Hyundai, the world's third biggest shipyard by revenue today, which produced its first large vessel – a tanker – in 1973, went on to become the biggest shipbuilder in the world within a decade.[34] This corporation exemplifies South Korean shipbuilding, which, close on Japan's heels, managed to climb to second place worldwide in the midst of the crisis in the 1980s. Thanks to huge government subsidies, South Korea was able to use the crisis years to rapidly expand its market share. In 1970, South Korea's share of global production was 0 per cent, in 1980, it was 4.3 per cent, and by the beginning of the twenty-first century, South Korea was the world's number one shipbuilding nation.[35] Ten years after South Korea began its ascent in the industry, the PRC embarked on its own accelerated catch-up race. In 1982, the government founded the China State Shipbuilding Corporation (CSSC) as part of a defence reform, soon concentrating on the construction of merchant vessels (the two largest shipping companies in China today are offspring of the original corporation).[36]

Faced with this overwhelming competition from Asia, Western European shipyards had little choice but to diversify into niche segments – such as the aforementioned cruise ship industry – or throw in the towel. Conventional tankers and large cargo ships had already become the domain of Japanese, Korean, and Chinese shipbuilders in the 1980s.[37] Only East European socialist states refused to follow this industry trend (for the time being). Poland and the non-aligned Yugoslavia, for instance, which relied on the foreign currency generated by this sector – among other things to repay the extensive foreign loans they had been receiving since the mid-1970s and that had been used to develop the export industry – continued to invest in shipbuilding and its expansion. To use the terms coined by David Harvey, the state socialist economies of Eastern Europe responded to their structural crisis by implementing both a "temporal fix" – deferring liabilities to the future –

and a "spatial fix" – tapping into new markets.[38] In Pula, despite years of stagnating demand, the number of people employed by the Uljanik shipyard peaked in the late 1980s, with over 8,000 people working for the company at the time – not to mention a few thousand *kooperanti* or subcontractors. At this point, the shipyards in Pula and Gdynia had no reason to be concerned about the increasingly rigorous competition rules that had become the new economic policy paradigm among the Eurocrats in Brussels and that stigmatized industrial policy, dismissing it as anachronistic. The different capacities that capitalist versus state socialist political economies had to adjust (or not) to an external shock could not be more apparent.

The Crisis and Failed Solutions of the 1970s

The global shock hit the two shipyards at a time when they had both just received massive investment to expand their capacities to enable them to participate in the growing market for tankers in particular. Poland's Five-Year Plan for 1971–5 included the largest investment the country had ever made in shipbuilding. As a result, on 30 October 1976, Gdynia's new dry dock (see figure 2.2) was opened to great fanfare, thus consolidating its position as Poland's most modern shipyard. At 380 metres long, 70 metres wide, and 8 metres deep, the massive dock was suitable for the construction of large container ships and general cargo vessels. With a dock of this size, Gdynia had the capacity to build ships measuring over 240 metres in length and with a deadweight tonnage of more than 100,000 DWT. Among these vessels were what were known as *marszaly* freighters, all named after Red Army Marshalls (with the *Marshall Budyonny* being the first), which Gdynia had been building for the Soviet Union since the early 1970s. At the same time, Gdynia Shipyard continued to produce small vessels (such as fishing boats), giving it a rather broad – one could say incoherent – production portfolio.

Much like Gdynia had done, Uljanik, too ventured into the field of mega ship and tanker production in the years leading up to the oil price shock. In 1970, the shipyard signed a contract with the major Norwegian shipping company Bergesen for the supply of a 265,000 DWT general cargo vessel measuring 330 metres in length – by far the largest ship Uljanik had built so far (see figure 2.3). The significance of this order, symbolizing as it did the pinnacle of shipbuilding in Pula, indeed in the whole of Yugoslavia, was portrayed in two documentary films made in 1972 showing the construction of the vessel: *Kolos s Jadrana* (The Colossus of the Adriatic) and *Berge Istra*.[39] Uljanik's director described these vessels, which were dubbed "mammoths," as the future of the

Figure 2.2. New dry dock in Gdynia, opened in 1976
Source: Courtesy of European Solidarity Centre, © Janusz Uklejewski.

Figure 2.3. *Berge Istra*
Source: Courtesy of Historical and Maritime Museum of Istria.

shipyard, saying this would enable it to finance itself going forward.⁴⁰ This claim would soon prove to be erroneous, when, from 1974, the shipyard found itself unable to sell its tankers without incurring crippling losses.⁴¹ The huge amount of money that had been invested into producing these massive ships – the lack of a dry dock the size of Gdynia's meant the ships had to be built in two separate halves that were then welded together – turned out to be a washout.⁴²

The downturn of the international market in the mid-1970s could not have come at a worse time for the two shipyards, as the political conditions could hardly have been less conducive to restructuring. In the wake of the 1970 workers' protests that shipyard workers were at the very forefront of, Poland's communist regime was keen not to lose the last shreds of its legitimacy; meanwhile, in 1974, Yugoslavia had just adopted a new constitution. The latter was a reaction to widespread dissatisfaction with the political status quo. In the years prior to this, this discontent had manifested itself in what was referred to as the Croatian Spring, while in other Yugoslav republics, it had led to liberal reform efforts that were initially met with repressive measures on the part of Tito. Consequently, neither government had much political capital at their disposal to expend on potentially painful economic reforms, not least because they still very much relied on the loyalty of the "working class."

Against this backdrop, the new and highly complex constitution Yugoslavia adopted in 1974 was to prove an unfavourable turning point. Overall, the constitution accelerated the country's process of decentralization and Yugoslavia took on a confederate character, giving the individual republics as well as the two autonomous provinces of Kosovo and Vojvodina a whole range of options that would enable them to block or prevent initiatives on the federal level. After Tito's death in 1980, the charismatic leadership component was gone, too. Consequently, the coordination problems brought about by the constitution were soon evident, as the consensus required for decisions to be taken in the federal bodies became more and more difficult to reach. Instead, the individual republics and provinces increasingly tended towards self-interest, a fact that made formulating and pushing through economic reforms all the more problematic.

The new constitution was followed by the 1976 "Law on Associated Labour."⁴³ This law was the pinnacle of the self-management that had already been initiated in the early 1950s, a principle introduced in all areas and at all levels of society. Companies were dissolved as legal entities – even the constitution no longer used the word "company/enterprise" (*poduzeće/preduzeće*). From that point on, production, management, services, and so on were to be organized by what were known as "basic organizations of associated labour" (*osnovna organizacija udruženog rada*, OOUR), which were legal entities representing the essence of self-management (each with their own workers' council). By 1980 there were already over 90,000 of these basic organizations. Business relations between the organizations were governed through what were known as "self-management agreements" (*samoupravni sporazum*). In order to stop the use of these agreements getting out of control, different basic organizations were able to create a "work organization" (*radna organizacija*, RO), which in turn could form an umbrella organization, referred to as a "complex organization of associated labour" (*složena organizacija udruženog rada*, SOUR). In any event, the law stipulated that key decisions – for example with whom business could be done and under what conditions or how revenues were to be distributed – must be made by the workers' council (*radnički savjet*) of the basic organization.[44] Even decisions about foreign trade relations were the prerogative of the work organizations.[45] In short: the grassroots were the ultimate authority (i.e., the bottom was at the top).

Uljanik was split into five work organizations based on the key areas of production. For example, there was an RO for shipbuilding and an RO for engine construction, as well as a total of eleven (and later twelve) basic organizations, each with their own workers' council, their own management, and their own heads. These basic organizations then established the Uljanik complex organization of associated labour, which served as a coordinating body and which outwardly behaved like a company, despite no longer having this status formally.[46] Karlo Bilić, the director of the Uljanik complex organization, dutifully extolled the enormous advantages of the "constitutional transformation [*transformacija*] of the collective."[47] But the shipyard's workforce was all too well aware of the shortcomings of the new system. In a poll conducted in 1982, the majority of the company's workers criticized the "fragmentation" of the business. In their view, "the socio-economic relations between the basic organizations were poorly defined and badly coordinated and were not well aligned with the objectives of the overarching organization (including the common product: the ships they built)."[48] An organizational theorist could not have summed this up better.

The small-scale developments observed in the shipyard also applied to the Yugoslav economy as a whole. According to Carl-Ulrik Schierup's analysis, the new organizational principles served only to reinforce the "bureaucratic perceptions" of the country's economic development. "Market criteria for judging the success of plants and enterprises, or the level of workers' wages, became increasingly irrelevant."[49] The only reason Uljanik and other large enterprises survived the next decade (and even then, it was touch and go) was because Yugoslavia's economic system provided financial relief through what were known as "soft budget constraints." As pointed out by Hungarian economist János Kornai, the socialist state acted "like an overall insurance company" taking on "all the moral hazards."[50] As a consequence, Kornai argues, "the attention of the firm's leaders is distracted from the shop floor and from the market to the offices of the bureaucracy where they may apply for help in case of financial trouble."[51] At almost exactly the same time, the director of Uljanik described the shipyard alluding to Kornai's rationale: "Under socialism we simply do not know how to wind up a system this size."[52]

Over the course of reform, the management lost its ability to run the shipyard. The capacity cuts that, as described above, shipyards in capitalist countries had to implement, were virtually impossible in a self-management system, even though in the mid-1970s, Yugoslav shipbuilding was in "deep water" and "shutdown" was looming, as reported by a local Pula newspaper.[53] Letting workers go, however, was simply not an option, explained Ivo Vrandečić, the head of the Yugoslav shipbuilders' association Jadranbrod, in a discussion with representatives of the metalworkers' union in 1976.[54] In other words, there was to be no shrinking of the bloated shipbuilding sector. Indeed, Uljanik even increased its workforce from around 6,000 in the mid-1970s to 8,124 by 1989. In this respect, Gdynia proved to be more flexible – in 1988, the Polish shipyard employed a total of 7,636 workers, which was 40 per cent less than ten years earlier.[55]

In the second half of the 1970s, the difficulties described above became serious, putting the continued existence of the shipyards at risk. In 1978, Gdynia found itself in the red for the first time; the shipyard came through this relatively unscathed, however, its losses, which ran to half a billion złoty, being borne by the state shipbuilding association.[56] The next few years saw Gdynia continue to operate at a loss, not least because it was forced to sell its ships for less than the production costs.[57] Although the Soviet Union was always a willing customer, trade with the country was based on barter and thus indirect payment, which after inflation deduction turned out to be less and less, as the Soviet Union was increasingly beset with its own economic problems. Uljanik's financial situation was certainly no better. In 1976, after a fire broke out on a tanker that was being built for a Nigerian customer, the shipyard was forced to request a government loan in a bid to avoid bankruptcy.[58] In 1980, Uljanik's bank accounts were frozen for several months because it could no longer meet its financial obligations, something that happened quite frequently over the following decade. The depreciation of the US dollar following the collapse of Bretton Woods exacerbated these problems, as, although the majority of supplier products from European countries had to be purchased in their own currency, ships were paid for in US dollars, which were now worth less in national currency. Yugoslavia let the dinar appreciate against the US dollar, which meant its trade deficit increased and the position of exporters deteriorated.[59]

Given how important shipbuilding was for both countries' economies, the governments in Warsaw and Belgrade were not going to just sit back and watch their biggest shipyards' reel from one crisis to the next. The shipyards were "finalizers" at the end of supply chains, thus also indirectly helping upstream companies to export. Moreover, they had a significant amount of social and cultural capital at their disposal: they symbolized successful industrialization in once "backward" regions, they were seen as incubators of socialist workers, and they represented the maritime orientations of both Poland and Yugoslavia. What is more, the two countries' governments feared the spirit of resistance that prevailed among the shipyard workers, something that the strikes in both Gdynia (1970) and Pula (1967) had borne witness to.[60]

Nevertheless, because of the different types of institutional conditions in the two countries, as well as the different export markets, the aid measures also varied. Poland was still a textbook example of a "centrally controlled" command economy with state ownership where sectoral ministries exercised direct control over the companies. In Yugoslavia, in contrast, with its system of self-management, the economic units enjoyed far more autonomy, and sectoral ministries were long a thing of the past. What is more, political responsibilities were so widely dispersed among the government bodies at the federal, republic, and municipal level that often no-one ultimately felt responsible. When Gdynia failed to meet its plan targets in 1979, the Polish government simply decided to replace the director – something that would not have been as easy in Yugoslavia. On the other hand, the Polish government did not have the systematic methods needed to be able to support a whole industry in crisis as a result of an external shock. Moreover, by this point, the oil crisis was impacting the national budget and the economy, as the Soviet Union had begun to demand higher prices for the oil it was selling to its allies. Much like all the other countries in the Eastern Bloc, the Polish government had no "Keynesian" stabilizers in its economic policy toolbox to help cushion the impact of the economic cycle.[61] Instead, it borrowed heavily from the West.

In Gdynia, management deficiencies and shortcomings in the organization of production were already in evidence in the late 1970s – before the 1980 strike wave and the mobilization of the Solidarność movement gave the operations management even more worry lines. The strikes in Gdynia began on 15 August 1980, just one day before the historical walkout at the Gdańsk shipyard, delaying the completion of ships that had already been ordered. In a report written at the end of September, the director of the Gdynia Paris Commune shipyard estimated the total number of working hours lost due to the strike to be 1.5 million.[62] Of all Polish shipbuilding companies, Gdynia Shipyard ended the year with the largest shortfall against the plan targets.[63] After another two bad financial years, in 1983 the company lost its creditworthiness.[64]

The Polish government tried to alleviate the crisis in shipbuilding by consolidating the industry. It created an association of the six biggest shipyards and their twenty-one most important suppliers, a body resembling Yugoslavia's bureaucratic superstructure Jadranbrod. In its first year, the industry association concentrated mainly on developing an anti-import program, which was to limit imports of the necessary equipment and upstream products for the construction of new ships to a minimum (in Yugoslavia, too, import substitution became a failed mantra for rescuing the shipyard). This strategy was, however, extremely unlikely to lead to big savings because only 30 per cent of the materials used in building a ship actually came from domestic suppliers.[65] The association also created a shared foreign currency account for its member companies so that they could draw on part of their export revenues to purchase the necessary upstream products from abroad.[66] This step towards corporate autonomy, however, served the interests of the individual companies only to a limited extent – in other words, only as long as they were making a loss.

In Yugoslavia, the governments in Belgrade and Zagreb were both reluctant to directly intervene in the shipyard's day-to-day operations. Micromanagement on the part of the state contradicted the basic ideas behind self-management, especially since this meant the government would have been in the firing line of any criticism. At the same time, however, the powers-that-be could hardly sit back and do nothing, given the seriousness of the crisis the industry was facing. From 1973 to 1978, the output of Yugoslavia's shipyards plummeted from a record two million DWT to as little as 236,000 DWT. For a country that was so heavily reliant on foreign currency revenues and where the shipbuilding sector had the highest export ratio of all industries, this development was cause for great concern. In response, the government began to develop a strategy to help rescue the shipbuilding industry involving the key stakeholders. This strategy was based on three pillars: identify new customers, finance the construction of new ships, and then focus more on selling to customers on the domestic market. Only when it came to the first of these objectives was there evidence of some success, and even then, this success was questionable from a financial perspective.[67]

Given how much of Yugoslavia's shipbuilding industry was concentrated in Croatia, it was only fitting that the government of this republic become actively involved in resolving these problems. This commitment began with a detailed report on "the problems of shipbuilding and shipping" presented by the industry ministry in January 1975.[68] The report opened with the observation that, in light of its 24,000-strong workforce as well as the more than 20,000 additional jobs provided by its subcontractors, shipbuilding in the Socialist Republic of Croatia deserved special attention. Moreover, the report continued, the shipbuilding industry in Yugoslavia had achieved the "highest foreign currency effect."[69] By way of contrast, the report focused on the problems the industry was facing: significant overcapacity on the world market, competition from Japanese shipyards, ships being sold at dumping prices, and the increasing subsidies Western governments was providing for their shipbuilding industry making it impossible for Yugoslavia to keep pace.[70] As a consequence, there were barely any new orders from Western customers. The only ray of hope was the Soviet Union, which managed to come through the global recession completely unscathed and which was expanding its oil extraction. That said, the Soviets did not pay in hard currency, which meant there was a limit to how attractive they were as customers.

The report also emphasized the declining competitiveness of Yugoslav shipbuilders owing to their outdated technology and lack of capital. Low efficiency was reflected in high prices: data from the late 1970s show that the price of Yugoslav ships was, on average, between 31 and 38 per cent higher than on the world market.[71] One of the reasons for these high prices was that upstream products from Yugoslavia were 50 per cent more expensive than comparable materials on the world market.[72] However, shipbuilders were politically forced or at least put under pressure to purchase as much as possible from domestic suppliers, irrespective of any business logic. The government tasked the export sectors to help non-competitive industries indirectly integrate into the world markets. Thus, the shipyards were, for example, required to use steel from a plant in Skopje, despite the fact that it was of inferior quality, often delivered late, and indeed more expensive than foreign steel.[73]

The government was aware of these problems, emphasizing that shipyards could "have only minimal influence over the price of their product ... because the price of the materials used in production was set primarily by state bodies and epitomized low productivity, poor organization, and outdated technology, all of which were then built into the ship."[74] Thus as "finalizers," the shipyards were at the head of a chain of inefficiency. A participant at one of the frequent meetings held to discuss the problems of the industry dryly commented that in shipbuilding "the conflicting interests of multiple domestic producers and domestic customers of these products collide."[75] It was not as if Yugoslavia's economic experts lacked an appreciation of the problems. What they lacked was the political courage to solve them.

This assessment of the situation and the political debates that took place at different levels produced one key proposal on how to solve the problems of Yugoslavia's shipbuilding industry. The shipyards were to increasingly gear their production towards domestic commercial shipping lines, which, up till that point, had been buying their vessels from abroad, where they were cheaper than at home. In June 1976, in the city of Piran on Slovenia's Adriatic coast, under the aegis of the Government of the Socialist Republic of Croatia, representatives of the country's shipyards and shipping companies along with government bodies signed a self-management agreement for the "construction of ships by domestic producers for domestic shipping companies."[76] What was known as the Piran Agreement became a source of permanent frustration – for the shipping companies the problem was a lack of financing to cover the additional costs for purchasing more expensive Yugoslav vessels, whereas for the shipyards it was the loss of foreign currency revenues and export subsidies that was the issue. The commercial banks refused to help.[77] Thus, by mid-1978 only five of the agreed forty-eight ships had actually been supplied to the domestic market. Against this background, it is hardly surprising that a second self-management agreement on the promotion of production for the domestic market, signed in 1981, also proved ineffective.

Thus, Yugoslavia continued to pursue an intensive foreign trade policy in order to find customers to buy its ships. With the aim of increasing exports, in the second half of the 1970s, the country dispatched trade delegations, along with a representative of Jadranbrod, to countries around the world, including Algeria, France, Iran, Cuba, India, the Soviet Union, Morocco, Uruguay, Venezuela, Colombia, Egypt, Syria, Somalia, Nigeria, Gabon, Congo, the Solomon Islands, the countries of the Arabian Peninsula, Romania, Bulgaria, and Turkey. These efforts were not in vain. Uljanik continued to build the majority of its ships for foreign customers, albeit frequently selling them at prices that did not even cover the costs. Of the fifty-four ships that Uljanik delivered in the 1980s, only eight were destined for Yugoslav customers. Some of the ships that were exported did however eventually end up, albeit via a circuitous route, in the national merchant fleet, because the Yugoslavian shipping companies leased the vessels from their owners in flag of convenience countries such as Panama and Liberia.

One specific problem was the availability of loans to finance the construction of new ships and export guarantees.[78] It was really only regional banks that were involved in shipbuilding; Riječka banka and Istarska banka, in particular, more or less acted as branch offices of the shipyards, which jeopardized their own liquidity.[79] In 1978, in order to provide relief to the industry, the government founded the Yugoslav Bank for International Economic Cooperation (Jugoslavenska banka za međunarodnu ekonomsku suradnju, or simply Jubmes) under federal legislation. This financial organization specialized in providing support for Yugoslav companies' international business, shipbuilding being the main beneficiary.[80] Jubmes refinanced export loans from commercial banks until the late 1980s, when, despite its special drawing rights from the reserves of the national bank, its coffers ran dry, too.[81]

In the end, it was not Yugoslavia that served as a putative lifeline for the industry, but rather the Soviet Union – although this lifeline would in fact end up pulling the shipyard down even further. For now largely unscathed by the global recession of the 1970s, the Soviets ramped up their oil production and foreign trade, meaning they needed new ships. Some companies placed orders with Yugoslav shipyards based on long-term bilateral trade agreements between the two countries.[82] In the 1980s, Uljanik produced twelve ships for the Soviet ship importer Sudoimport – more than one-fifth of the shipyard's entire production for that decade.[83] One of the flagship projects within this program was a train ferry that could transport 108 carriages.[84] There was just one small problem, as the head of the industry association Jadranbrod, Ivo Vrandečić, was forced to admit: not even "a magician" could make any money doing business with the Soviets.[85] They paid either in (non-convertible) roubles or in kind and, moreover, they were all too aware that they had the upper hand in business negotiations in this field because the demand for ships was so low. In 1977, for instance, they insisted that, in future, the full price of the ship would be paid only on delivery, whereas previously, 50 per cent of the total cost was due while construction was underway. Responsibility for financing the construction of new ships thus shifted entirely to the shipyard, which increased their borrowing requirements even more.[86] To make matters worse, deliveries to the USSR were not eligible for export subsidies, unlike ships exported to hard-currency countries where the state granted price subsidies of between 20 and 33 per cent. In an interview in 1979, Uljanik's director Karlo Bilić lamented, "It is not unusual for us to build ships that make no profit at all."[87] Capacity utilization thus came at a high price.

Despite the aforementioned financial difficulties, both shipyards continued to build ships, albeit in decreasing numbers and at a loss. That being said, Gdynia's crisis was less of a threat to its existence than Pula's, as the mechanism of soft budget constraints meant that the shipyard's losses were more or less automatically covered by the state shipbuilding association. The decline in production, based on tonnage built, was also less dramatic as this was the time that the giant *marszaly* were being built in Gdynia. The most important aspect in this context, however, was that Soviet demand served as a buffer against the global crisis – a unifying element in the history of the two shipyards, albeit with different financial implications for each. This meant a reversal of the strong orientation towards the Western markets observed in the preceding decade.

On the Brink of Financial Ruin

The 1980s proved to be even more difficult for the two shipyards than the decade of the oil crises. The global recession that hit in 1981 and the anti-inflation policy implemented by the United States, which brought strong appreciation of the US dollar in its wake as well as rising interest rates worldwide, meant that global demand for ships remained very low and the annual production of shipyards around the world failed to reach the 1969 level again.[88] The Western industrialized countries were still busy reducing the overcapacity they had created before 1974. The only exception here was South Korea, a new player in the industry that, over the coming years, acquired an ever-stronger position. Given the unexpectedly long duration of the crisis, the willingness of the Polish and Yugoslav governments to help the beleaguered shipbuilding firms (as well as companies in other industries) was increasingly pushing their financial limits.

Both countries were heavily indebted to Western creditors, with Poland's financial obligations to Western banks amounting to around 25.5 billion US dollars in 1980. When, beginning in autumn 1979, the Federal Reserve continued to increase the interest rates in a bid to tackle inflation in the United States, these loans became even more difficult for Poland to refinance. At the same time, the country slid into a deep recession that was to last three long years. In 1981 alone, Poland's GDP declined by 10 per cent.[89] Then, when it imposed martial law, Western creditors temporarily refrained from granting the country new loans.[90] In order to enable the country to repay its foreign debts, Poland's government tightened foreign exchange control and froze company accounts, which meant that it was virtually impossible for Gdynia Shipyard to import components from the West. The situation in Yugoslavia was not much better: in 1983, the government's foreign debts amounted to almost 23 billion US dollars, which was just short of one-third of its national product and double the export revenues it received in convertible foreign currency.[91] In order to fill the depleted foreign reserves again, the government introduced reforms that allowed it to tap into banks' and companies' foreign revenues, making the lives of companies that were reliant on imports for their production all the more difficult.[92] In Yugoslavia, too, the phase of economic growth was over, and, in 1983, the country recorded a decline in its economic output for the first time since the war. Moreover, Yugoslavia faced soaring inflation rates, reaching almost 170 per cent in 1987 and hitting the 240 per cent mark a year later.[93] Poland also recorded two-digit inflation rates throughout the decade.

In light of this highly problematic situation, it was clear that the policy of "soft budget constraints" was going to become all the more difficult to finance – not least because it was also a cause of general malaise. The government in Warsaw did very little in the way of trying to overcome the prevailing economic challenges. The regime was preoccupied with mere survival, having violently suppressed the popular Solidarność movement and imposed martial law in December 1981. Although the government under General Wojciech Jaruzelski was aware that economic reforms were essential to its survival, it had neither the political nor the financial capital to improve the economic situation in the country.[94] At the same time, it was reluctant to risk further alienating its industrial workforce by implementing a structural reform of shipbuilding and other heavy industry. This thankless task was left for the post-communist government that came to power in 1989.

In Yugoslavia, meanwhile, the death of Tito in 1980 uncovered the structural weaknesses and inherent obstacles and impasses in the system that was his legacy, a system that was at once decentralized and autocratic with collective leadership at the national level. With Yugoslavia having lost the ultimate arbitrator in its hypercomplex political system, it proved nigh on impossible to pursue a consistent policy of reform. Given the growing economic problems and social crises the country was facing, however, such a policy was vital. The orthodox members of the party and the reformers cancelled each other out, and it became harder and harder to bridge the divergent interests of the individual parts of the federation. According to a report by the CIA, export companies also contributed to the weak enforcement of political decisions. They "barely complied with the requirement ... to transfer their foreign currency to the central bank and repatriate the income [generated abroad]."[95]

Although the two shipyards managed to carry on building ships, the economic rationale behind this was increasingly questionable. Gdynia faced a sharp decline in orders – in particular from domestic companies – supplying only three ships in 1983 and a paltry two in 1985. In an attempt to alleviate its loss, the shipyard attempted to enter a new market segment: the construction of passenger ships. The first vessel of this type, the *Stena Germanica*, was scheduled to be delivered in 1982 but in the end did not actually reach its Swedish customer until 1987. Consequently, the Stena Shipping Company promptly cancelled the order for the remaining three ferries.[96] The penalty for delayed delivery was a colossal 17 million US dollars, a sum that exceeded the total value of the shipyard's exports in 1985.[97] An audit report produced by the Polish Court of Auditors stated that "despite tax relief and privileges, the profits generated by the shipyard did not cover its liabilities."[98] Gdynia's debts amounted to 31 billion złoty, resulting in the company having to pay 3.9 million per day in penalties to the national bank for arrears on its interest payments.[99] It was only thanks to a state aid package approved in 1984 that Gdynia was able continue production.[100]

The shipbuilding industry serves as an example of the overall ambivalence of Poland's economic policy in the second half of the 1980s as the country headed towards payment default on its debts towards Western creditors. The government was gradually running out of funds to finance the losses of its state-owned enterprises and thus had no choice but to grant them more economic freedoms, which, in turn, increasingly exposed the problems of the individual companies. Initially, the shipbuilding companies did indeed try to stand on their own two feet and integrate into the world market, only to return to the shelter of the state after the aforementioned setbacks.[101] The state-owned enterprises discovered the downside of the concepts "autonomy" and "individual responsibility," which, in the years that followed, increasingly took the form of an implicit threat from the state that one day it would refuse to provide a safety net. This posturing took on a more tangible shape when, in 1988, the government announced the closure of the famous Lenin shipyard in Gdańsk, which at that time was building more ships than any other shipyard in Poland, intimating that Gdynia would be next in line. At the watershed of 1989, the end of the proud history of

Polish shipbuilding appeared to be imminent. In 1990, Gdynia Shipyard built no more than four ships, including two trawlers, which meant they were far from utilizing their full capacity.

Uljanik found itself in a similarly precarious situation in the 1980s. In both 1980 and 1981, the shipyard's bank account was frozen for months on end because it could not meet its liabilities, which at the time amounted to two billion dinars plus another 20 million US dollars, even reaching the point where the solvency of its main creditor, Riječka banka, was at risk.[102] At the same time, the young shipyard director, Karlo Radolović – according to economic historian Michael Palairet a tough and capable manager[103] – faced the challenge of making the fragmented company capable of reform. Here, the director's many years' experience in the shipyard's operations stood him in good stead. Radolović's most important achievement was streamlining decision-making processes. He did this by shifting the power vertically to the central coordination level of the shipyard, making it clear to the basic organization of associated labour that it was to accept the decisions of this central level even if they were not in line with what was laid down in the Law on Associated Labour. He replaced the existing five procurement departments for each work organization (*radna organizacija*) with a single procurement system.[104] A referendum conducted in the individual basic organizations in 1985 approved a more centralized structure for the shipyard, which then gradually began to resemble a normal company again. This "bottom-up" decision essentially paved the way for the recentralization of companies – a step that was taken four years later under a new corporate law in Yugoslavia. Radolović managed to get the workers' council to agree to overtime and the cancellation of workers' leave in an attempt to ensure the ships were completed on time, thus avoiding penalty payments – an achievement that, considering the system of self-management, was no mean feat.[105]

Based on the recommendations of a Danish business consulting firm, the management of the shipyard began to introduce upgrading measures that aimed at producing smaller but highly specialized vessels, thus opening up a market niche for itself.[106] Not only did the company purchase new cranes and machinery as well as licences from Western companies (e.g., for engine construction); more importantly, it introduced IT-based design and production.[107] Uljanik had acquired its first IBM computers at the end of the 1970s and, in 1985, upgraded these to 16-megabyte central processing units, which enabled more than one hundred terminals to be connected. This degree of processing power allowed the shipyard to use computer-aided design and computer-aided manufacturing, ushering in the "era of computerization" at this early stage.[108] The renown of Uljanik's IT department, which had programmed its own shipbuilding software, was demonstrated by the decision by the organizers of the 1984 Olympic Games in Sarajevo to commission the shipyard to install the IT systems required to monitor the competitions and races.

The shipyard earned an excellent reputation, especially for its complex multipurpose vessels, which, to use the company's own words, were "a clear trademark of Croatian shipbuilding and a competitive export article."[109] However, the mainstay of production was still tankers – for oil and chemicals –, primarily for customers in the Soviet Union. Uljanik's director also found a creative solution to the problem of ships that were not collected by the customers who had ordered them. This solution was Uljanik Plovidba, a shipping company founded by the shipyard in 1986.[110] This move had a dual rationale, as explained by director Karlo Radolović in an interview. First, Uljanik hoped to be able to cut the losses it had incurred from customers failing to pick up these ships by using them itself. Second, the shipyard could use the ships it owned as collateral for loans it needed to finance the necessary equipment and upstream products and to build new ships.[111] One thing the shipyard certainly did not lack was entrepreneurial creativity. Uljanik even established a department for financial *inženjering*, which, in 1992, was lauded by Radolović as "the future of Pula and Istria."[112]

These measures led to a temporary revival. In November 1986, the shipyard announced that it would be employing 1,250 new workers over the next five years with a view to increasing its capacity to nine ships per year.[113] In an article on Uljanik's export potential, the national daily newspaper *Vjesnik* sang Uljanik's praises, even calling the company "one of the most important shipyards in the world."[114] In January 1987, *Glas Istre* referred to the "Rolls-Royce of shipbuilding" that was being produced by Pula's shipyard.[115] The quality of the ships built by Uljanik did indeed enjoy an international reputation, although this barely reflected in its revenues. Just ten months after these success stories, Uljanik's accounts were frozen yet again – this time for almost two years.[116] The immediate trigger here was the "Sudan program," which had since become a "Sudan problem," as one Croatian daily with an appreciation of irony reported.[117] This episode clearly demonstrated the cost of foreign policy that saw trade as a means of fostering bilateral relations and made exports to non-aligned states so symbolically charged that any notions of economic profit were a mere afterthought.

In the late 1970s, Sudan, which was a non-aligned country and thus a friend of Yugoslavia, commissioned the shipyard in Pula to build three ships. The Sudanese shipping company failed, however, to pay the purchase price of over 14 million US dollars. Although years of negotiations and high-level diplomatic talks, as well as interventions by the Yugoslav national government, did bring about some creative proposals from the Sudanese (who offered to pay in cotton rather than cash), no solution was found. As the Yugoslav shipbuilding association pointed out at the time, Sudan's failure to pay pushed Uljanik "into a situation of total illiquidity, leaving the shipyard with no means of continuing its operations."[118] On top of this, Uljanik had already accumulated considerable debt. The shipyard owed almost 20 billion dinars in unpaid taxes and duties as well as over 130 billion dinars to the banks, not to mention its 52 million US dollars in debt to foreign creditors and suppliers. The Belgrade daily newspaper *Večernje novosti* even asked, "Is the record-holding company about to go bankrupt?"[119] The rest of Yugoslavia's shipbuilding industry was faring no better, despite the fact that, going by the tonnage on the order books, the country had climbed to third place worldwide. Nevertheless, the shipyards remained huge loss-makers. Uljanik had plenty of work but no money, lamented the company's finance director in autumn 1989.[120]

Continuing to Muddle Through in the Post-Socialist World

In 1988, demand for new ships reached its lowest point. From then on, trade grew much faster than global GDP.[121] This meant that, after years of being in their death throes, shipbuilders were enjoying an expanding market once again. In 2004, the record value of 36 million GRT (gross registered tonnage, a measure of the internal volume of commercial ships) that had been reached in 1975 was achieved again; in 2011, the total global production of 103 million GRT set a new record in shipbuilding history.[122] The upturn of the world market arrived at precisely the right time for the shipyards, especially from a psychological point of view. The increased demand provided them with renewed prospects, filled their order books, and helped them to more or less fully utilize their capacity. They at least managed to create the impression of having a glorious period of growth ahead of them, which gave legitimacy to their continued requests for government support.

To ensure they were properly equipped for the transition, in October 1990, the shipyard management in Pula commissioned Price Waterhouse to develop a future strategy – the end of communist rule made ideological blinkers obsolete. The international consulting company was unsparing in its analysis of the status quo: "Until recently, the company's main aims were employment and social welfare, while profitability was of secondary importance."[123] That said, the analysis did at least identify a few strengths,[124] including the shipyard's many years of experience building specialized ships, an excellent reputation for the quality of its products, its favourable location (in Istria, shipbuilding could continue throughout the winter), "strong informal relations," as well as the "government's willingness to support the shipbuilding industry." Yet the picture painted by the far longer list of weaknesses was of a company that was lacking in organization and targets, never delivered ships on time, relied on outdated technology and antiquated equipment, and had no marketing to speak of. This was a firm whose bookkeeping left much to be desired, that suffered from poor work discipline and high overhead costs, did not care about occupational safety, had very little money, and operated at a significant loss. The costly advice from Price Waterhouse was simple: in the future, the shipyard had to focus on actually earning money.

This was certainly not a bad idea, but was far easier said than done, given that the economic conditions in the country were worsening with each passing day. Croatia's economy was in freefall, shrinking by an average of 10.2 per cent a year between 1989 and 1993.[125] Although the fighting did not spread to Istria during the Croatian War of Independence (1991–5), the shipyard in Uljanik did not remain unscathed, with foreign investors avoiding the crumbling Yugoslavia and thus also Croatia. What is more, a number of men who had worked for the shipyard were also enlisted in the Croatian army.

Compared to Croatia, Poland's political transition went without a hitch, but the economic ramifications were just as devastating. In 1990 and 1991, the country's gross domestic product plummeted by around 18 per cent and the number of unemployed skyrocketed to over 2.3 million.[126] Owing to its political significance, heavy industry was propped up by the government but still had to shed huge numbers of jobs. In both countries, the high inflation made future planning incredibly difficult for shipyards. To make matters worse, the Polish shipbuilders temporarily lost their most loyal foreign customer of recent years – the Soviet Union, which collapsed in 1991. In his article about three Polish shipyards (Gdańsk, Szczecin, and Gdynia), Preston Keat summed up the situation:

> Each of the three shipyards faced generally similar challenges. They all were too big, lacked focus, were unaccustomed to hard budget constraints, and, to complicate matters further, they had all just lost their most important customer, the USSR. They all had very broad product portfolios – they produced too many different things. They also maintained a variety of 'social assets', including hospitals, housing, and vacation hotels.[127]

And virtually the same could be said of Croatian shipbuilding at the time. It was crystal clear to the managers in both Gdynia and Pula that their shipyards' only hope of survival was to establish themselves in market niches, thus resolving their liquidity problem. This was the only way they would be able to finance the upgrading of their plants and new shipbuilding projects. Because this restructuring process was proceeding at such a sluggish pace both shipyards received huge amounts of assistance from the government in order to get the changes required underway or at least to ride the waves of the initial years of transformation. Despite their commitment to capitalism, the new powers-that-be did not in fact hang the shipyards out to dry – at least not at that moment. It was not only for sentimental reasons that shipbuilding was close to the governments' hearts. More importantly, it was because of its role as the biggest employer in the manufacturing sector along the coast – in Croatia even more so than in Poland.

In the 1990s, it initially seemed as though the paths taken by our two case study countries might actually diverge significantly. While for Uljanik, virtually the entire decade seemed to be about sheer survival, the shipyard in Gdynia blossomed into an outstanding European example of successful restructuring. Indeed, the contrast with the shipyard in Gdańsk was particularly illuminating. As Preston Keat argues, the former Lenin shipyard relied far too heavily on its importance as a political symbol and the birthplace of Solidarność. Instead, Gdynia (and the shipyard in Szczecin) became a flagship project for the reform of heavy industry.[128] Gdynia profited from substantial debt relief when, in 1993, its creditors accepted a 50 per cent debt cut totalling 240 million US dollars. One-third of this was shouldered by the banks, a further third by the government, and the rest was covered by more than seven hundred suppliers. The remaining debt was converted into shares for the 280 largest creditors. The substantial outstanding balance with the Polish Social Insurance Institution (ZUS) was written off, resulting in the publication of an article in the local press that ironically referred to the social security provider as an "institution at your service."[129] The debt cancellation left 40 per cent of Gdynia Shipyard in private hands.[130]

The government, which was still the majority shareholder, pushed ahead with the restructuring of the shipyard, despite the initial strong resistance from the workers (something that was far less visible to the outside world). As Keat observes, the trade union leaders in the shipyards gradually managed to "popularize the restructuring programme among workers."[131] The reforms helped the shipyard turn itself around and in 1995, its employment figures exceeded those of 1989 and real wages more than doubled. Even productivity increased dramatically, evidenced, among other things, by the fact that the time the company needed to build a medium-sized container ship went from sixteen months in 1992 to twelve months in 1994. By 1998, the total ship construction time (from start to finish) was as little as five months.[132] The dry dock inherited from the socialist era, the more streamlined portfolio, and the craftsmanship and engineering skills of the shipyard's employees led – aided by a favourable international market environment – to a temporary upswing. In 1995, Gdynia was making a profit again – a remarkable turnaround, if we consider that, in 1993, the shipyard had recorded the third-highest loss of all Polish companies.[133]

Shortly after this, the shipyard began a short spell as a private enterprise. In 1998, when the state treasury transferred a further 12 per cent of its shares to the shipyard's employees, it lost its position as the largest single shareholder. "We are taking things into our own hands," announced the shipyard's new director, Janusz Szlanta (appointed in 1997), acknowledging this shift in ownership.[134] Gdynia's CEO, Szlanta, even embarked on an expansionary path, taking over the Gdańsk shipyard that had declared bankruptcy in 1998. Here, Szlanta also had his eye on a vast holding that belonged to the Gdańsk company, a space where, similar to Western shipyard sites, office complexes and residential buildings later developed.[135] The humiliation of being bought up by local competitors was something the 3. Maj shipyard in Rijeka was also to experience (in the hands of Uljanik), although here, too, this was a last resort and not part of a strategy to give birth to a "national champion." The post-socialist governments had neither the vision required nor the necessary financial resources to have pursued such an industrial policy.

In 1999, Szlanta even expressed an interest in purchasing two shipyards in Finland – meaning Gdynia was filled with optimism. One important reason for the improved business situation was the management's decision to cut its product range. While in the past, Gdynia had built virtually every type of ship, from fishing boats to huge tankers and freighters to research vessels, it then concentrated on containers, ro-ro ferries, and vehicle transporters. In 2000, the shipyard produced fifteen ships: six container ships, six bulk carriers, two LPG tankers, and a car transporter (the latter, Crystal Ray, seen in figure 2.4, was the last huge vessel produced in Gdynia).[136] In the years that followed, production in Gdynia was to shift even more towards the construction of car transporter ships, a segment in which the shipyard was able to carve out an important position on the world market – in direct competition with Uljanik.

Figure 2.4. *Crystal Ray*
Source: Photograph by Alf van Beem, 21 February 2005, Wikimedia Commons, https://commons.wikimedia.org/wiki/File:Crystal_Ray_IMO_9210440_photo-1.jpg.

As we can see, one important factor in Gdynia's success was the shipyard's director Janusz Szlanta, who had built quite a reputation by the late 1990s. Szlanta was held in high regard not only because of his good relations with the government but also for his role in industrial revitalization. In one of the interviews we carried out, a former engineer recalled, "Of course he stimulated production hugely. We even managed to build up to eighteen ships a year. He literally inflated the shipyard" (E.P.). The annual tonnage built by the shipyard increased to over 300,000 GRT again. In 2000, the Association of Polish Managers selected Szlanta as Manager of the Year.[137] At the awards ceremony, Szlanta said,

> This position is our very own achievement. We didn't fight for it in the capital, we didn't woo politicians for their support. Our arena is international competition. Without state support – and in an industry that is a symbol of inertia and lack of prospects at that – we managed to build up a company that is a European leader. The Europeans have the greatest of respect for Polish shipbuilding; they are even a little afraid of us.[138]

The jury did not object to the fact that, at the time, the state treasury still owned 22 per cent of the shipyard's shares and that the company had been able to achieve such success only thanks to the generous state debt relief. They were much more inclined to see its expansion as a shining example of how Polish companies could become respectable players on the European market. This new self-confidence was the reason Szlanta won the award. Full order books enabled the shipyard to justify its requests for continued government support even though the majority of its shareholders were private. As in Pula, the irony here was that the company's success did not reduce the demands on the state but in fact served only to underpin them. Dry docks working to capacity and the construction of huge container ships that can hold 2,700 containers as well as transporters for 6,000 cars gave the impression that Gdynia really was a very successful business.

That said, full capacity utilization was not synonymous with profitability – productivity growth flattened, profits fell from 114 million in 1998 to four million złoty (at the time just over a million euros) and a black zero in 2001, and even that was achieved only with difficulty. And Gdynia was certainly in good company; according to an analysis, in the year 2000, European shipbuilders' average operational margins were coming in at 0 per cent.[139] On top of this, there was still the problem of dependence on loans to finance the building of new ships, which made the shipyard heavily reliant on banks.

In the early 1990s, Uljanik faced similar problems to Gdynia. The shipyard suffered almost permanently from a lack of liquidity – a fate it shared with the majority of other companies in Croatia – and had massive difficulties accessing credit. The government at the time was frugal and attempted to keep the national budget as balanced as the war permitted.[140] From 1990 to 1992, Uljanik was unable to land a single new contract, leaving it with no advance payments to improve its cashflow. The order books of the five biggest Croatian shipyards together had depleted from a peak of sixty ships (2 million DWT) in 1986 to seventeen vessels (0.7 million DWT) in 1992.[141] From the end of 1989 to summer 1992, Uljanik's losses accumulated to 3.5 billion Croatian dinars (272 million US dollars).[142] When, in preparation for its impending partial privatization, Uljanik conducted a company valuation, it turned out that the 2,700 apartments it owned across the city were its most valuable asset (see map 4.1). In August 1991, Uljanik Standard, the division that managed the apartments, was valued at 130 million German marks, while the shipbuilding division (*Brodogradilište*) was worth just 68 million marks – 70 per cent of its book value.[143] The decision to sell the apartments to the tenants in the mid-1990s, albeit at a modest price, was a crucial way for the shipyard to acquire short-term liquidity.

A very real consequence of Uljanik's lack of funds was the recurring problem of paying wages on time. On 19 October 1992, the trade unions decided for the first time ever to call a strike over wage arrears. The strike saw more than 5,000 (of 6,400) workers down tools for four hours.[144] When the workers threatened to strike for a second time, guarantees from the government, a local bank and the Istrian Chamber of Commerce meant the outstanding wages were paid three days later.[145] In the years that followed, work stoppages were a relatively common occurrence.

In addition to the shipyard's failure to pay wages, inflation also exacerbated the social situation that workers had to endure. Salaries had fallen to a pitiful level in real terms. In 1992, the annual price increase in Croatia was over 600 per cent, and in 1993 it even reached as much as 1,600 per cent. In April 1994, around 450 workers went on strike in protest against the low wages (the hourly wage was the equivalent of approximately 1.25 US dollars, compared to the average 20 US dollars per hour in Western European shipbuilding) – two days later the strike came to an end, with the management agreeing to a 200 per cent wage increase.[146] The workers were not the only ones to suffer from the shipyard's notorious financial difficulties – Uljanik's business partners felt the impact, too. The power companies were forced to switch off the shipyard's electricity supply on several occasions because it had failed to pay its bills. There were even cases of Uljanik employees being refused treatment at hospitals because the shipyard had failed to transfer outstanding health insurance contributions. Given the increasingly unattractive employment conditions, Uljanik also struggled to find qualified staff. There were, for example, not enough ship mechanics on the local labour market to fulfil new orders. Moreover, there were fewer and fewer young people looking to pursue a career in the shipbuilding industry, reducing the demand for relevant specialist training. This trend, which had already started under socialism, now intensified. The shipyard was no longer at the heart of the community.[147]

It was not until the end of the Croatian War in 1995 that the situation began to improve and the government found itself in a position to focus more on economic policy without doing so at the expense of social factors. During the war, the patriotic mobilization had obscured the social costs of economic transformation, at least in public debate. The government had limited its efforts to "palliative measures" in order to keep the shipyards alive despite their chronic cashflow problems.[148] Then, in August 1995, the government finally passed a law that allowed the *sanacija* (restructuring) of large state enterprises.[149]

October of the same year saw the launch of a program for the reorganization of the country's five largest shipyards. Based on recommendations from the German consulting company Rödl & Partner, the program's main objective was to reduce the shipyards' debt burden and in so doing help them to recover their liquidity.[150] Uljanik's debts with the Riječka banka (the shipyard's principal bank), the government pension and health care funds, the energy provider, and the Ministry of Finance totalled 144 million US dollars.[151] The government wrote off the lion's share of these debts, pushing domestic creditors, all of which were state owned, to accept a conversion of their receivables into company shares.[152]

These measures did little to change the desolate situation of the Croatian shipyards, however. There had been barely any investment in equipment and machinery since the 1980s and this had clearly affected productivity, which was catastrophic in international comparison. According to the deputy minister for economic affairs, who was responsible for the restructuring program, Croatian shipbuilders needed between 64 and 88 man-hours to construct one CGT (compensated gross ton; a measure for international comparison in ship production that takes its complexity into consideration), whereas German shipyards needed only 32 man-hours and Japan as few as 25. At the same time, the number of effective working hours per year in the Croatian shipyards was as little as 900–1,000, compared with the EU average of 1,500–1,700 and as many as 2,100 hours in Japan.[153] A second, related problem was the low number of incoming orders. In 1993, the capacity utilization of Croatia's shipyards was down to just 65 per cent and in 1994, they received just six new orders, while the Poles signed contracts for thirty-seven new vessels in the same year. The press lamented that Croatia, which had once been a leading shipbuilding nation, had toppled from third place worldwide (based on tonnage in the order books) to thirteenth place in just a few short years.[154]

Once its debts had been reduced, Uljanik, in collaboration with government experts, set itself the ambitious goal of reaching Germany's 1995 productivity level of around 32 man-hours for one CGT and 1,630 effective annual working hours by the end of the millennium.[155] This implied a considerable reduction in the workforce from around 5,000 employees in mid-1995 at the start of the *sanacija* to about 3,300 a year later. These staff cutbacks were a sensitive issue, especially for the trade unions, although they ultimately agreed to the government's plan, being well aware that the alternative was the closure of the shipyard.

Employee representatives nevertheless managed to negotiate a social plan. The majority of the workers who had lost their jobs were given early retirement or an invalidity pension, one-third were laid off with a one-off redundancy payment of 10,000 kuna (around 2,700 German marks at the time).[156]

Although the provision of a social safety net to mitigate the impact of these layoffs is reminiscent of the structural changes carried out in Western Europe in the 1980s, the differences between the two situations must not be overlooked. The redundancy payments were not flanked by extensive government-funded upgrading of production equipment or publicly financed staff retraining measures. Those who had been laid off in Pula in 1995 – or a couple of years earlier in Gdynia – were often left with nothing. This applied in particular to female employees, who were disproportionately impacted because they tended to work in the "unproductive" parts of the company, such as the canteen, holiday home for employees, or administration. At both local and company level, Poland's and Croatia's transformation thus lacked the future-oriented components of Western European structural change, which at first thought may seem paradoxical given that the post-communist governments promised their people a better future under the new free-market economy. Proactive government policy, however, works on the premise that the coffers are full, something that was not the case in Poland and Croatia due to hyperinflation, high foreign debt and huge problems with the major domestic companies. The second premise was the existence of relevant institutions such as the social partnership, which had not yet been created.

What Uljanik could rely on, however, was the promise of government subsidies to the tune of 15 million US dollars until 1999, which went into upgrading the shipyard's machinery. Under the de facto protection of the government, Uljanik used these years to better organize its production. As a result, by 1996, it was the first industrial enterprise in Croatia to be awarded ISO 9001 certification for quality management (Gdynia then managed the same under Szlanta).[157] In 1999, the managing director of Uljanik confidently announced that the shipyard had overcome the crisis and in 2000, for the first time in many years, the company recorded an annual profit.[158] The fact that, for the first time, in April 2000, a private bank (*Privredna banka*) had agreed to finance three new projects signified the end to Uljanik's worst years and the return to full creditworthiness.[159] In September 2000, Uljanik was officially released from the state restructuring program. Production now concentrated on multipurpose vessels, car and truck transporters, and ro-ro ships. At the end of 2001, Uljanik had fourteen ships on its order books. The livestock carrier *Becrux*, which was soon to be delivered to the Italian shipping company Grimaldi, symbolized the restored confidence and technical prowess of the shipyard. *Becrux* was one of the biggest and most modern vessels of its kind in the world. It comprised nine decks and could transport 14,000 cows or 75,000 sheep across huge distances to places as far away as Australia, New Zealand, or the Middle East. The almost 180-metre-long livestock carrier was equipped with its own desalination plant, which had the capacity to produce 600,000 litres of fresh water per day.

All of a sudden, Uljanik became something of a role model for Croatia, while news from the other big shipyards unsettled both the public and the government. In 1998, their cumulative annual losses amounted to a shocking 600 million US dollars.[160] The restructuring program had certainly not benefitted these shipyards to the same extent and, at the turn of the millennium, they needed another hefty capital injection. The deputy minister for economic affairs, Milan Čuvalo, who oversaw the restructuring program, accused the shipyard management, the trade unions, and the municipalities where the shipyards were located of blocking the real restructuring efforts, saying they had missed the opportunity for reform under state protection.[161] The directors had joined forces with the trade unions in order to keep employment high. Finance Minister Borislav Škegro demanded that the shipyards sack their directors for their failure in the restructuring efforts, calling upon them to appoint managers "who did not constantly harass the state but instead demonstrated the will to restructure."[162] The trade unions and the managers vehemently rejected these accusations from Zagreb and Uljanik's director Karlo Radolović expressed his solidarity by writing a furious letter to the government blaming the ministry experts for the sorry state of Croatian shipbuilding and accusing them of having out-of-touch ideas.[163] In the end, the government granted new loan guarantees and threw even more money at the problem. Uljanik received one hundred million German marks as a non-repayable subsidy for technological upgrades.[164]

In with the EU, Out with the Shipyards

The decline of world trade after the global financial crisis of 2008/2009 heralded the end, albeit with a slight delay, of the recent global shipbuilding boom that had lasted more than two decades. Uljanik and Gdynia had both temporarily profited from this boom – but their Asian competitors had gained the most. Whereas in the rest of the world, the tonnage produced between 1991 and 2012 remained relatively stable, in

South Korea, the GRT rose more than six-fold, in Japan it tripled, and Chinese production continued to grow exponentially anyway, increasing from 0.5 million GRT to 35 million in 2010 (see also figure 1.2).[165] By the end of the boom phase, the top three East Asian producers accounted for more than 80 per cent of the output of merchant vessels. In 2000, South Korea overtook Japan as the world's biggest producer of merchant ships, while China outstripped both of its Asian competitors in 2010. One of the distinctive features of Chinese shipbuilding, which was also a significant advantage, was that around one-third of the ships built were bought by domestic customers – a level of domestic demand that Croatia and Poland could only dream of. Another important trait of the Chinese shipbuilding industry is the dominance of state-owned enterprises: the two most important companies, the China State Shipbuilding Corporation and the China Shipbuilding Industry Corporation, which account for 70 per cent of production, are "state-owned enterprises under the direct supervision of the State Council."[166] This state backing helps them to master the challenges of the world market – while the EU waves the flag of fair competition.

The dominance of the three Asian nations is also evident at the company level. The fifteen biggest shipbuilding groups are all in Asia: eight in South Korea, six in the PRC, and one in Japan. The largest European producer, Germany's Meyer shipyards, headquartered in Papenburg, ranks only in thirty-eighth place.[167] At the start of the 2010s, the biggest individual producers were the Korean companies Daewoo and Hyundai, each of which completed more than sixty ships in 2012, followed by another Korean enterprise, Samsung, which managed to build fifty ships in the same year. The production of cruise ships is the only segment that is still the domain of the Europeans.[168]

Although Uljanik did not enter the cruise ship market right at the turn of the millennium, it did not make any more negative headlines either. For a few years, it even made a profit and, when it did find itself operating at a small loss, the government was quick to offer a "helping hand," for instance through loans or export guarantees. Zagreb saw no urgent need for action, which can partly be explained by the fact that, of all the Croatian shipyards, Uljanik was seen as a success story, even though it had not yet been privatized. Uljanik even helped bail out the heavily indebted repair shipyard Viktor Lenac in Rijeka.

Uljanik's recipe for success in the first decade of the twenty-first century was to concentrate on building complex ships designed to specification. For example, for customers from Russia, Azerbaijan, and later also Turkmenistan, Uljanik began to construct specialized vessels to transport road vehicles and railroad cars across the Caspian Sea. In this context, Croatian foreign policy served as a door opener, drawing orders into the country, as a former adviser to the foreign minister at the time told us during an interview. Uljanik became a global leader in this market niche.[169] Reduced costs and shorter construction periods for new ships made the company competitive again.[170] When, in January 2006, Karlo Radolović finally stepped down from the position of director (he remained a board member), passing the baton to Anton Brajković, it seemed as though Uljanik had successfully weathered the post-socialist transformation. In 2006, the shipyard celebrated its 150th anniversary, publishing a splendidly illustrated book with pictures of 250 ships it had built since the Second World War. At the time, there were fifteen ships with a total value of almost one billion US dollars on the order books – this was certainly a reason to be optimistic.[171]

Around this time the clouds above Stocznia Gdynia had begun to darken again. The insolvency of the shipyard in Szczecin in 2002 had prompted the banks to demand more loan guarantees from Gdynia, too. The appreciation of the złoty, triggered by the prospect of EU accession, and the decline in demand for dry bulk carriers at the beginning of the decade made it increasingly difficult to finance production. When in 2002, Gdynia's director Janusz Szlanta, who just a few years earlier had confidently announced that he was planning to take the fate of the shipyard into his own hands, asked the state for help, this was widely interpreted as "the end of the maritime industry's liberal policy."[172] Some commentators were even of the opinion that the case of Gdynia showed that "the Polish People's Republic was still alive."[173] It goes without saying that Szlanta's call for help pertained mainly to the government guarantees that were needed to make it easier to finance contracts – returning control to the state was certainly far from his mind.

However, the social democrats, who were in government at the time, took a different approach. The government was not prepared to make concessions until a representative of the state treasury had taken on the chairmanship of the supervisory board and Szlanta had stepped down from his position on the board. At the same time, the shipyard's debts continued to accumulate to such an extent that it was clear it would not be able to survive without government assistance. In 2004, the company faced "renationalization," a process that, once feared, now clearly had no alternative (in the next chapter we will analyse the meandering and permeable boundary between the state and the private in more detail).[174] After a cash injection from the government had increased its share capital, the state treasury owned 52 per cent of the company's shares. The state was able to present itself as a saviour, while the once celebrated CEO, Szlanta, was now being hounded by the courts. The finance minister and government agency for industrial development drew up a plan for a large shipbuilding holding that was to consolidate all the loss-making companies in the industry. In fact, this idea had already been mooted under the Polish People's Republic.

The 2008 European Commission report, which concluded that state aid granted to Gdynia Shipyard was illegal, includes an unsparing analysis of management failings. In a restructuring plan developed in 2004, the management of the Gdynia Shipyard Group placed all the blame for its financial troubles on external factors, "such as Asian competition, strong position of the zloty against the dollar, [and] a difficult situation in ship financing." However, the EU competition authorities disputed this, claiming that the management assumed unacceptable "design, technological, financial and commercial risks" and failed to safeguard the group against the appreciation of the złoty.[175] Its financial difficulties "led to arrears in settling public and civil law obligations and wages, shortages in material supply, slowdowns in production processes, cost increase (labour consumption, penalties), and delays in fulfilment of contracts." The Commission criticized the strategy of filling the order books with no regard for potential losses that, much like with Uljanik, could already be foreseen when the contract was signed. Capacity utilization should not be an end in itself, at least not at the risk of going heavily into the red.[176]

In 2004, Gdynia Shipyard was still proud of the fact that, based on the volume of orders placed, it ranked 13th in the world. This was certainly not an accurate measure of business success, however. The drop in the number of workers employed directly in shipbuilding from 6,249 in late 2004 to 4,611 in summer 2007 was not enough for the shipyard to catch up with its competitors in terms of productivity. Even at a time when global demand for new ships was booming, the two attempts made by the government (in 2006 and 2007) to find an investor or buyer for what was then the biggest shipyard in Europe were in vain.

In Gdynia – and since Uljanik stopped production, in Pula, too – the commonly held opinion is that the EU is to blame for the demise of the shipbuilding industry. And like many conspiracy theories, this does contain a grain of truth.[177] Accession to the European Union brought great upheaval for the new member states in general and for shipbuilding in particular. Neither Stocznia Gdynia nor Uljanik managed to survive for long once their respective homelands had become members of the EU. A mere five years under EU conditions was enough to irretrievably destroy the local shipbuilding ecosystem and the interrelationship between the state and business. Uljanik and Stocznia Gdynia were not the first shipyards to experience this: the large Neptun shipyard in

Rostock stopped building ships in as early as 1991, a year after German reunification, which was also the first step of eastward expansion of the EU (then European Community). Neptun initially became part of the Vulkan shipyard in Bremen, which went on to declare bankruptcy itself in 1996.

Owing to the paradigm of competition policy and equal treatment on the single market, EU accession implied the definitive end of the protection the Polish and Croatian governments had given to their shipyards for so long for the purpose of shielding them from the vagaries of the world market. The consequences of non-compliance with EU competition rules are something Gdynia Shipyard was soon to discover first-hand. When the shipyard was essentially renationalized after suffering major losses, the Polish government's financial support had already overstepped the EU requirements. Once part of the EU, a state was no longer allowed to make a sovereign decision to pursue a policy of "soft budget constraints." In 2005, the European Commission opened a formal investigation examining the aid Gdynia had received as it had its doubts "that any of the conditions for the aid to be approved as restructuring aid had been met."[178] The investigation revealed that, between May 2004 and June 2007, the Polish government had granted five guarantees for the construction of new ships, written off debts on four separate occasions (with a total value of 30 million złoty), provided fresh equity amounting to 40 million złoty and given the shipyard a total of 65 million złoty in loans. According to the calculations by the Commission, the total aid provided up to September 2007 amounted to 5 billion złoty (approx. 1.434 billion euros at the time).[179]

In the Commission's opinion, these payments were in breach of European competition and state aid rules, not least because the Polish government was unable to provide any valid reasoning as to the purpose of the subsidies. The restructuring plan proposed by the strategic investor ISD Polska did not convince Brussels either, given that it required further subsidies. The Commission bluntly countered that the 2012 productivity target set down in the strategy – 37 man-hours/CGT – fell far short of the levels efficient European shipyards had already attained (20 man-hours/CGT).[180] According to the Commission, the proposed investment strategy was more of "a collection of small upgrading projects without the global vision which would appear necessary to modernize facilities many of which date from the 1970s."[181]

The Commission's decision that "to restore the status ex ante, the State aid has to be recovered," was the death knell of the shipyard in Gdynia. Just one month later, on 5 December 2008, the Polish government passed a special act on the liquidation of the Gdynia and Szczecin shipyards. The only consolation for the workers was a social plan that had been negotiated with the trade unions. Depending on years of service, the employees who were laid off received redundancy pay of between 20,000 and 49,000 złoty (approx. 5,300 to 13,000 euros).[182] In an ironic twist to the story, it was the Gdańsk shipyard of all places – split off from Gdynia again since 2006 – that passed the review conducted by the Commission. The company's subsidized privatization was found to comply with the EU's state aid rules. However, the Commission demanded that the Gdańsk shipyard reduce its capacity – its workforce was to be cut to around 2,000 employees and parts of the site were to be handed over to other companies (today, among other things, steel towers for wind farms are manufactured on the shipyard's territory).[183]

Croatia already experienced the stringency of the EU's state aid rules and the Commission's less than enthusiastic attitude towards shipbuilding during its accession talks. Indeed, for Croatia, the massive amount of state aid became one of the biggest obstacles to accession (along with deficits in the rule of law and cooperation with the International Criminal Tribunal for the former Yugoslavia in The Hague). As had already been established by the Organisation for Economic Co-operation and Development (OECD), Croatian shipyards received higher subsidies than those in other EU countries.[184] By signing the Stabilization and Association Agreement in 2001, Croatia had already undertaken an obligation to restructure its shipyards within four years of the agreement entering into force. Moreover, the EU required that the shipyards move forward with their privatization by 2013 – otherwise all state aid from 2006 onward would have to be paid back.[185] Nevertheless when Croatia joined the EU in 2013, thanks to years of government subsidies, the country was still home to the same five big shipyards of forty years previously, while two-thirds of European shipbuilding companies had closed.[186]

The accession process moved the difficulties faced by Croatia's shipbuilders back into the public eye, sparking renewed efforts by the government to restructure industry. Just as in the past, the government in Zagreb and the majority of Croatia's general public saw shipbuilding as a strategic industry. In 2007, the shipyards employed more than 16,000 people and accounted for 12 per cent of Croatia's exports (ships and shipbuilding materials were the largest product group in the country's export structure).[187] At the same time, given the broad support for EU accession, the EU's requirements meant that the attitude of both the government and the public towards shipbuilding began to shift. The times of this industry being able to rely on unconditional support from the government and the undivided sympathy of the population were over. In its place was resistance to the apparently insatiable appetite of shipbuilding companies for fresh injections of state money. In 2007, for instance, a boom year in global shipbuilding, Croatian taxpayers forked out 430 million euros for the shipyards, 53 million of which was direct subsidies, while the biggest Croatian shipyards recorded losses of 146 million euros and their accumulated debts amounted to more than 3 per cent of the country's economic output.[188] Their productivity also continued to leave a lot to be desired. In 2010, their output was the same as the Danish shipyards, despite having four times the workforce.[189]

On the eve of EU accession, the government opened its coffers for the last time, but now state aid was tied to clear restructuring conditions. The government promised the shipyards in Split, Rijeka, Trogir, and Kraljevica more than 1.8 billion euros (13.4 billion kuna) in restructuring aid, 1.3 billion of which was assumed debt.[190] Some twenty years after privatization had been mentioned for the first time – a goal that had cropped up repeatedly over the years, only to disappear as quickly as it had come – Zagreb was now serious about restructuring through privatization. The trade unions were not exactly thrilled; their fear was that the shipyards could be closed, but their resistance was ultimately futile.[191] As required by the EU, the restructuring resulted in a substantial reduction in production capacity amounting to more than one-fifth, necessitating a 25-per cent reduction in the overall workforce across the country's five biggest shipyards.[192]

Uljanik did not appear on the list of companies to be restructured as the management had successfully convinced the government and the European Commission that the shipyard was sufficiently well positioned to tackle its problems single-handedly. From 2008 to 2010, Uljanik, recording profits of between 23 and 76 million kuna a year, believed it did not need any further restructuring aid and could thus avoid the strict regulations.[193] However, what the management failed to take into account here was the ramifications of the global financial and economic crisis: four of the biggest ships that were scheduled to be completed in 2008/9 ended up in the shipyard's own shipping company, Uljanik Plovidba, which acted as a kind of "bad bank" for cancelled or rejected ships.[194]

So it seemed Uljanik could not avoid privatization either. However, due to a lack of potential buyers and given the trade unions' opposition to foreign investors, the path to privatization was somewhat unorthodox.[195] With the government's blessing, in July 2012, more than 1.5 million shares were sold at a 20 per cent discount to almost 7,200 individuals, the majority of whom were current or former shipyard employees. Together, they purchased 46 per cent of the company's stock, unexpectedly making them the main shareholder twenty years after the end of "social ownership" and the interim stage of nationalization. Of the other shareholders, the biggest were now government pension funds, commercial banks, and insurance companies, all of which clearly had close links to the government. Even so, Zagreb still proudly announced, "We have now successfully privatized the Uljanik shipyard."[196] And the European Commission agreed.[197] In 2013, Uljanik even bought its long-standing competitor, the 3. Maj shipyard in Rijeka, for the symbolic sum of one Croatian kuna, privatizing this company, too, and fulfilling another of the EU's requirements into the bargain. Moreover, by buying 3. Maj, Uljanik gained access to the extensive restructuring aid that Zagreb had granted the Rijeka shipyard.

Everything seemed to be going well. The order books began to fill up again, as the international crisis in shipbuilding began to ease from 2014 (with the global output of new ships having fallen from over 100 million GRT in 2011 to just over 60 million GRT). The decline in demand for ships had resulted in renewed reduction in capacity. After 2009, the number of shipyards building big ships worldwide had more than halved, from 372 to just 120 in 2021,[198] causing some of Uljanik's competitors (like Gdynia) to leave the market, too. Uljanik's decision to specialize in ships built to customer specification proved to be a good move, at least in the short term. In 2013, the Uljanik Group recorded profits totalling 103 million kuna (around 15 million euros). From 2013 to 2017, they delivered nineteen ships, not to mention the numerous platforms for natural gas production and small dredgers.[199]

On top of this, several highly acclaimed new projects were started, including for example, the world's first luxury ship for cruises to the polar regions – this was to be Uljanik's first cruise ship. Another of the shipyard's flagship projects was the world's most powerful cutter suction dredger. But Uljanik would never complete these ships – following the company's bankruptcy, the cruise liner *Scenic Eclipse*, which was to be built for an Australian customer, was completed by the shipyard in Rijeka, now divested again (nowadays you can book outrageously expensive voyages on the vessel). The fate of the dredger *Willem van Rubroeck*, commissioned by a Luxemburg company, will feature prominently at – and as – the end of this book.[200] Uljanik lacked the technical capacity and the financial means to complete these ambitious and challenging projects on time, which is why these much-vaunted contracts only ended up causing more problems because of the penalties the shipyard had to pay.

In 2016, the market situation deteriorated again. Incoming orders plummeted worldwide and remained at a historic low in the years that followed.[201] Shipbuilding is after all, as stated by the OECD, a cyclical industry that frequently faces extreme fluctuations in both directions. Uljanik had failed to adjust to these ups and downs. The reprieve from the toils of transformation proved to be short-lived. At the end of 2017, the shareholders, the government and indeed the general public expressed surprise that the financial year had closed with massive losses of 1.8 billion kuna (almost 250 million euros).[202] The apparent successes of the previous years were based on a growing mountain of debt resulting from the fact that, since privatization, the costs of financing the construction of new ships had increased dramatically because lenders now demanded different risk premiums.[203] What is more, the management, who received bonuses for each new order, concluded too many contracts, which the shipyard then struggled to complete on time, thus incurring penalties. There were media reports about questionable business transactions entered into by the executive board headed by Uljanik's former financial director, and now CEO, Gianni Rossanda, as well as about inflated bonuses and costly consultant contracts – at times even with themselves.[204] In January 2018, the workers did not receive their monthly wages on time and suppliers too had to wait for payment. The management received a state guarantee for a loan in the amount of 96 million euros, which was intended to cover operating costs. The shipyard narrowly avoided strike action when the loan was approved by the EU Commission.[205]

In early 2018, a ballpark figure of at least 400 million euros was bandied around as the amount of restructuring aid the shipyard needed. Yet, unlike in the past, this time the government was not prepared to come to the rescue – and indeed was not authorized to because of the EU rules. Moreover, the management had failed to present a plausible restructuring plan that would have enabled the state to grant aid anyway. Instead, the shipyard began a desperate search for a so-called "strategic investor" – a truly mythical figure – which soon ran aground. First, the only person to express interest was a shady Croatian businessman with whom the management board already signed a takeover agreement but who then failed to come through and invest. Another Croatian entrepreneur then entered onto the scene, an individual who already owned the privatized shipyard in Split but who was unable (or unwilling) to come up with the 300 million euro asking price for the company. These ostensible investors were not prepared to invest their own money, instead expecting to receive subsidies.

In this horizon of expectations, there was not a lot of difference between these capitalists and the workers protesting against the government's reservations regarding financial support. The workers in

Pula organized demonstrations, downed tools, and, initially at least, were able to count on public, and indeed local government support for their actions. In summer 2018, the bridging funds were running out and the workers in Uljanik once again waited in vain for their wages. This time, things came to a head and there was a full-blown strike, both in Pula and in the 3. Maj shipyard in Rijeka. The workers called for the management's resignation and for the government to provide bailout funds to secure the survival of the two ailing shipyards. In August 2018, the Uljanik workers organized a protest march in the centre of Zagreb. At the time, they were still receiving the minimum wage, paid from an emergency fund set up by the government. But many of them had simply had enough and hundreds decided to quit their jobs – to take up employment at the nearby Fincantieri shipyard in Monfalcone in Italy or just to move to Germany like tens of thousands of other labour migrants had done since Croatia joined the EU. And just like that the mobilization of protest was over. In March 2019, the district court in Pazin opened insolvency proceedings (*stečaj*) against Uljanik, putting an end, at least temporarily, to Pula's 163-year shipbuilding tradition.

Afterlife(s)

In his highly influential book *Capitalism, Socialism and Democracy*, published in 1942, Austrian-born economist Joseph Schumpeter popularizes the term "creative destruction." Schumpeter writes,

> Capitalism, then, is by nature a form or method of economic change and not only never is but never can be stationary.... The fundamental impulse that sets and keeps the capitalist engine in motion comes from the new consumers' goods, the new methods of production or transportation, the new markets, the new forms of industrial organization that capitalist enterprise creates.

According to Schumpeter, the process of industrial mutation took a course that "incessantly revolutionizes the economic structure *from within*, incessantly destroying the old one, incessantly creating a new one." He considered creative destruction a constituent part of capitalism.[206] Without the introduction of new technologies and organizational principles as well as the entry of new companies, growth cannot be sustained. Capitalism will continue to forge inexorably ahead, provided nothing puts a stop to it and there are no monopolies to prevent innovation. In this process, uncompetitive companies are replaced by the more competitive ones.

It could be argued that the large-scale devastation of industrial capacity as state socialism was dismantled across Eastern Europe was more destructive than it was creative, given that, in many cases, nothing new emerged from what was destroyed. But here, it rather depends on the time frame – building something new atop the ruins of old industries can take a lot longer than originally anticipated. The economic recovery of the countries of Eastern Europe is testimony to this. It is precisely for this reason that historicization is so crucial. The legacy of a relatively good infrastructure, a well-educated and well-trained population with a highly qualified workforce, an industrial work ethic and institutions that embedded industry in the wider community helped some post-socialist countries to rebuild their industries after the dramatic downturn of the early 1990s. Here, the proximity of the prosperous Western European markets, especially Germany's, the influx of foreign investment and EU aid as well as integration into international supply chains and the European single market were all crucial elements – but without an industrial legacy, these factors would never have been effective.

As already mentioned, the big Stocznia Gdynia shipyard closed its gates in 2009 after the EU forced the Polish government to demand the return of its subsidies. At the time, many people in Gdynia refused to accept this as the end of their proud shipbuilding tradition. And, indeed, just a year later, the public limited company Crist began to use the former Stocznia site. On its homepage, Crist proudly describes itself (incidentally in English rather than Polish) as follows:

> We steadily develop and adjust our offer to changing market demands.... CRIST belongs to the group of companies distinguishing themselves through innovativeness, niche products and the organization of the supply chain. The company cooperates with Clients from Poland, Germany, Norway, Denmark, Finland, Iceland, France, Belgium, The Netherlands and Scotland.[207]

What began in 1993 as a small firm with a modest consignment of four fishing boats has become a real success story – at least for the time being. In 2017, Crist supplied the Finnish shipping company FinFerries with the hybrid battery-powered ferry *Elektra*, which won two international awards.[208] The company also builds hulls and other parts for cruise ships. With around 1,500 employees, Crist is far smaller than the original shipyard, but the company provides job opportunities for many engineers and skilled workers who used to work at Stocznia Gdynia.

Crist is, however, not the only example of life after death in the (post-) socialist shipbuilding industry. The Neptun shipyard in Rostock, which apparently went under following a botched privatization attempt, was then taken over by cruise ship builder Meyer shipyard in 1997, which then restarted production at a new site nearby in 2001.[209] The new Neptun shipyard mainly builds units for river cruise ships – another booming market segment.[210] What these attempts at renewal have in common is that they maintain a shipbuilding ecosystem with specialized engineers, skilled workers, infrastructures, supply chains, training institutions, and governments that ensure macroeconomic stability, as well as solvent banks that can lend money for production, and lastly opportunities for free trade within the EU. With both Gdynia and Rostock, the decisive factor was possibly the absence of fierce competition with tourism investors over the exploitation of a beautiful stretch of coast. In contrast with the clear blue Adriatic in Istria, the Warnow and Vistula estuaries will never be among the most attractive destinations for sun-worshipping, sea-loving tourists.

In Istria, on the other hand, the rumour has been circulating for years that powerful local economic interests were out to deliberately destroy or at least relocate Uljanik in order to free up the picturesque Bay of Pula for more profitable tourism purposes. Such suspicions were already expressed in August 2007 by the head of the trade union for Istria and Kvarner at the time, Rajko Kutlača. The latter accused the mayor of Pula of hypocrisy, claiming that representatives of the dominant regional political party, the Istrian Democratic Assembly (IDS), wanted Uljanik out of the bay. Putative investor Danko Končar's ultimately failed attempt to invest in the shipyard in 2018 further fuelled such speculations, as Končar had already purchased the concession for the tourist development of the northern side of the bay years ago, where he had planned to build a yacht marina along with a hotel (though so far this has been slow to materialize). Any phoenix wishing to arise from the industrial ashes of Uljanik would therefore have to be prepared for some fierce competition over this attractive stretch of coast.

Given that Croatia lacks some of the essential ingredients for even a modest industrial revival – including banks that are willing to invest and larger-scale upstream suppliers – and the country is entirely reliant on tourism, a development resembling that of Gdynia or Rostock seems highly unlikely at this stage. That said, there is currently also a lack of wealthy investors that are capable of transforming a site that has been used for military and industrial purposes for over 160 years into a tourist area. In the defunct shipyard, a slow process of decay has thus begun. The giant cranes (*dizalice*) of Uljanik are still and silent, but at least at night they are illuminated in a festival of colour.

3 A Safe Haven? The Role of the State in the Transformation

"Family photographs and half empty teacups" – this is the scene described to us by a marine engineer who had been allowed to accompany the insolvency administrator as they walked through the grounds of Gdynia Shipyard in 2010. "Everyone left their jobs in the belief that they would be coming back, as if they were just taking a couple of months off before returning to work." This expectation was not unfounded. After all, there was a precedent: in 1988, the last communist prime minister Mieczysław Rakowski, the eternal enemy of Polish dock workers, had attempted to close down the nearby Gdańsk shipyard – one of the cradles of Solidarność – but to no avail. Since then, the majority of workers had remained convinced that shipbuilding on the Baltic coast was untouchable. Even twenty years later, the workers in Gdynia's production halls saw this nationalized company as a safe haven and believed the politicians' lip service, still convinced they were all in the same boat. Indeed, back in 2009, the government was certainly inclined to protect the industry; but their hands were tied. All anyone could do was hope that the European Commission would soften its position or forget the whole debacle. As we have already seen, this hope turned out to be delusional.

"What would your father do if you suddenly turned up with no money to live off and asked for help?" quizzed the hairdresser of one of the authors on a visit to Pula in 2019. This rhetorical question came quite unexpected. The conversation had actually been about the Uljanik shipyard, which had recently gone bankrupt. Mario, the hairdresser, did not even attempt to conceal his inability to comprehend the Croatian government's (in)actions, which could not bring itself to rescue the shipyard this time. At the same time, his expectation is also an expression of a long-lasting tradition of state paternalism built on the interrelationship between the shipyard and the Socialist Republic of Croatia and later what was an independent Croatian state. How often had the government already had to pull out the stops to save the shipyard in Pula from its chronic illiquidity? Each new government bailout package further reinforced the expectation of the management and the workers that the state would throw the shipyard a lifebuoy during times of crisis – whether in the troubled waters of late socialism in the 1980s, during the unnavigable transition crisis in the 1990s, or on the stormy seas of the 2008–9 global financial crisis.

Both of the conversations described above made the research team, educated as it was in the paradigm of "neoliberal hegemony" in Eastern Europe, prick up its ears.[1] The history of the two shipyards as corporate entities is not really in keeping with the master narrative of entire societies being thoroughly privatized,[2] of ruthless shock therapy,[3] and of the downgrading of the working class.[4] Furnished with this expectation from the literature on the hegemony of neoliberalism, during our field research, we stumbled over phenomena such as rescue packages, renationalization, and politicians serving as CEOs that, initially at least, we did not know what to make of. A comparison with other European countries shows that shipbuilders in Poland and Croatia were not the only ones to have expectations of the state. A large-scale study conducted by the Sociological Research Institute (SOFI) in Göttingen in the 1980s showed similarly high expectations among employees of West German shipyards.[5] As historian Bo Stråth argues in his analysis of structural change in the European shipbuilding industry, in times of crisis, it was more the government than the management that the dock workers saw as having a duty of care.[6]

The shipbuilding industry constitutes a very specific context when it comes to the relationship between the state and industry. First, the shipyards traditionally have close ties with the national defence industry, which is why governments are notoriously reserved when it comes to foreign owners. Second, the longer time-to-market for big merchant ships as well as the high debt ratio typical for new constructions imply an almost incalculable risk for the shipyards and, since the 1970s, have often meant they have no choice but to secure government guarantees for financing – and the same applies in Western Europe or East Asia. Does this mean that the shipbuilding industry could be an exception within the neoliberal hegemony, according to which the state stays out of corporate decisions?

The close ties between the shipbuilding industry and the state, which can be seen the world over, do not, however, provide enough of an explanation for the variations in the public-private partnerships in different countries. On the contrary, in fact, within this ubiquitous symbiosis, there are significant temporal and regional differences. Everything suggests that, despite globalization, it is not only the industry that determines the conditions of production, but the nation state – in other words, the respective political economy. After all, the hairdresser in Pula is not the only one to think that way: Croatia is one of those EU countries where a particularly large share of the population endorses government intervention. According to the European Values Survey, more than half of the Croatian respondents were even in favour of more government ownership – in Poland, this value was actually slightly higher, at 55 per cent.[7] Bearing this in mind, we do not consider these two shipyards to be "deviant cases"[8] in a sector that was, given its special status, able to swim against the current. Instead, we prefer to see the shipyards as an impetus, steering us towards an interpretation of transformation that takes the diversity of possible paths seriously. With the help of the shipyards, we can tilt the image of transformation from a neoliberal monoculture towards multiple capitalisms, which change depending on the perspective they are observed from.

Looking at the changing role of the state through the prism of the shipyards captures path dependencies of post-socialist economic activity and counter-movements to rhetorical market self-regulation. In this context, a double movement comprising commitment to reform and safeguards against the market comes to light – not dissimilar to the mechanism Karl Polanyi described in his *Great Transformation*. With a focus on the state as a counter-force vector, we uncover the optical illusion of transformation. Here we choose not to concentrate on the well-known neoliberal reform rhetoric underlined by Mitchell Orenstein and Hilary Appel in their concept of "competitive signaling,"[9] but instead shine a spotlight on how reforms came to nothing and on the reactions to such failures. To this end, we analyse the meandering path of relations between the state and the shipyards, a path that is marked by both ruptures and surprising continuities. While the previous chapters provided abundant evidence of government attempts to protect the industry against the global market, this chapter addresses the contradictions in which the state became embroiled in the process. We will first focus on the swing of the Polanyian pendulum, as exemplified by the boundary line between companies and government authorities. We then look at the reconfiguration of the state sphere and, finally, we discuss the shift in the shipyards' strategies of legitimation towards political decision-makers.

The Shipyard's Porous Boundaries

"The factory gate was here," the erstwhile documentalist at Gdynia Shipyard announced in a meaningful manner, pointing to a roundabout around which the cars belonging to the managerial personnel are parked. Next to this is the office building that employees dub the "Aquarium." There is no longer any trace of the crowds of workers who once gathered here at shift change. The building is now home to a dozen small companies, the documentalist informs us, which is why there is no longer a main entrance. The disappearance of the factory gate is symbolically charged: it shows how the boundaries of the former company have changed. During a process that has been ongoing since the 1980s, what was once a juggernaut of an organization has become a fragmented collection of smaller enterprises. But had the factory walls not already signified but an illusion of clear boundaries between an "inside" and an "outside" before? The holistic nature of the centrally organized socialist company already went hand in hand with pronounced geographic disparity: the shipyard's sanatorium, for example, was located in Wieżyca in Kashubia; its sports club was in Bałtyk in Mały Kack, south of the city; and the apartments that belonged to the company's cooperative were distributed throughout Gdynia. The situation in Pula was no different: although the striking ramparts of the former Austrian armoury created a physical barrier around the shipyard, the latter still owned premises that were spread across the city.

A look at the final phase of state socialism shows that, during the course of transformation, it was not just the company's physical boundary that became volatile. In fact, in as early as 1980, a series of strikes in Poland – which shortly after this culminated in the establishment of Solidarność – called for the end of state paternalism, which ultimately meant the company's vertical relations were to be severed too. The 1981 Companies Act, which essentially defined companies as "independent, self-financing, and self-managed" reflected this commitment to reform. What this meant was that companies' boundaries were already up for debate under state socialism, and the expectations of the different actors were far from homogenous.

Although "the state" and "the system" were perceived as clear enemies of Solidarność, in the 1980s, resistance soon developed against economic reforms that reduced state actors to the role of a "founding body." Many of the shipyard's employees saw autonomy and self-reliance – terms that masked the governing party's expectations that companies should finance themselves from their own profits – as a Damocles sword over the reform-minded yet unprofitable companies.[10] Faced with the high staff turnover, gaps in labour standards, delays in the construction of thirteen ships, and poor worker morale, the boards of the Paris Commune Shipyard felt ill-prepared to break away from the state budget. An analysis by the "work committees" of the governing Polish United Workers' Party (PZPR) stated, "Under these conditions, the implementation of reform was stopped and restricted."[11] After martial law was imposed in 1981, this countermovement prevailed and the autonomy postulated by Solidarność remained fictional, at least for the time being. The imagined factory walls, shutting out the overbearing state, were soon torn down again; the physical factory gate marked the beginning and end of the working day for the employees but not the shipyard's boundary line. It was impossible to say with certainty where the state ended and the company began. The public and the private spheres overlapped, just as the external and the internal realm of the company did not constitute binary opposites under socialism.

Following a botched contract for the Swedish shipping company Stena in 1983, the situation in Gdynia had become unbearable. On top of current liabilities amounting to four billion złoty and loan debt of over 31 billion, the shipyard was also burdened with penalty payments imposed by the national bank for long overdue interest. According to the Supreme Court of Audit at the time, "This mountain of debt was impossible to pay back."[12] The shipyard was on the brink of bankruptcy, a concept that did not even exist in a planned economy. In 1984, the state, which had founded the company and modernized it at great expense in the 1960s and 1970s, stepped in one more time.

This intervention was, however, quite different from Western European practices. While countries such as Germany, France, or Sweden used subsidies to ensure that structural change was conducted in a – at least theoretically – socially responsible manner as well as to promote technological innovation in shipbuilding and to reduce capacity,[13] the socialist state put up a protective shield to prevent any potential restructuring. It was only thanks to the government decision in March 1984 to provide financial aid to the shipyards that the Paris Commune yard in Gdynia was able to survive.[14] The rescue package included accelerated debt write-off, bank loans at preferential terms, and subsidies to enable the completion of orders with foreign shipping companies. Despite the government's intention to take a more rigorous approach to insolvency law and what the national bank considered an insufficient restructuring plan on the part of the shipyard, the Ministry for the Steel and Machine Construction Industry still managed to persuade the national bank to approve more loans. This was an almost textbook example of what János Kornai has described as "soft credit," which served to "assist firms in substantial and chronic financial trouble without real hope of repaying the debt."[15]

Loans with barely any conditions were not the only element of soft budget constraints used in the case of the Paris Commune Shipyard. "Subsidies to cover losses," "subsidies to compensate for the profitability of the sale," and "item-specific subsidies" were just a few of the different types of government assistance that improved the shipyard's balance sheets in the late 1980s. What is more, these subsidies softly and silently tied the state budget to the company's accounts – the exact opposite of the demands the International Monetary Fund (IMF) had made of the highly indebted Poland. Kornai describes these various types of subsidies and aid, but also taxes and tax breaks, that the socialist state combined as "an intricate network of redistribution composed of 'many channels of income flow.' The central budget takes money from the firms on a hundred different grounds and gives it to them on another hundred."[16] In other words, the line between the company's budget and the state budget became overrun by a jungle of subsidies and transfers. One consequence of this blurring of budgetary boundaries was the ongoing collective liability for companies' failure. This practice could work only provided that the social collective was either not asked in the first place or a public consensus was reached to the effect that the shipbuilding industry was of such fundamental importance that the sacrifice by the community was justified.

The shipyard's reliance on this support from the state impacted its autonomy. In 1983, following the imposition of martial law in Poland in December 1981, the government drew up a list of 1,392 "companies of fundamental importance for the national economy" that included the shipyards. Contrary to what was included in the self-management reform act, in these companies it was the relevant government ministry rather than the workers' council that appointed their director. Moreover, these companies were still required to present its long-term plans to the ministry. A year later, the Ministry for the Steel and Machine Construction Industry extended its powers of intervention even further by appointing members for the supervisory boards of the six largest shipyards "to oversee compliance with the social and economic national plan and the government policy line."[17] Once again, the state reached deep into the company – the factory gate was easily surmounted, despite all the attempts at reform in the 1980s.

In both Yugoslavia and Pula, the subsidy jungle surrounding the Polish shipyards served as a cautionary example. The Yugoslav state, however, exercised more restraint towards companies in the name of "self-management." Here, the government did not own the companies, society did (whatever that meant). Unlike in the Polish People's Republic, in Yugoslavia, economic policymakers tended, as far as possible, to respect the integrity of the company accounts in their choice of measures. Meticulous care was taken to draw a clear distinction between the company (*poduzeće*) and its financial cycle, on the one hand, and public institutions (*ustanove*), on the other, and company contributions to the latter were recorded as parafiscal duties.[18] Since the late 1950s, the League of Communists has also refrained from interfering in the running of companies.[19] Not even when faced with the downturn of the world economy in the mid-1970s did the government abandon their self-restraint. Rather than eliminate this sharp distinction, the Croatian parliament, the Sabor, chose to deepen it: the parliament resolved to return the taxes as well as the contributions Uljanik had paid into the federal fund set up to help the poorer regions of Yugoslavia.[20] Moreover, the "Platform for the Solution of the Problems of Shipbuilding and Shipping" recommended that the city of Pula exempt Uljanik from local taxes.[21] Such tax reductions constituted the third way pursued by Yugoslavia between the laissez-faire approach of the free market and direct subsidies; they kept the illusion alive that companies did not require a state protection mechanism.

From this vantage point, the representatives of the Yugoslav shipbuilding industry were critical of the high subsidies granted by Western industrial states, claiming that they distorted competition.[22] Yugoslavia feared it would be squeezed out of the market by a subsidy race. Under this pressure, state actors moved away from the sharp distinctions between government and company accounts so they could act as an intermediary between companies and the global demand for ships. In 1981 and 1987, the only way to sustain Uljanik's production was through capital injections from the government and loans from the Yugoslav national bank as well as the regional banks Riječka banka and Istarska banka. Loans were distributed according to the ideologically sacrosanct maxim of solidarity. Economist Vladimir Gligorov argues that, in the self-management economy, solidarity was supposed to act as a corrective to the invisible hand of the market: "It provides sound economic foundations for self-management and, at the same time, avoids certain unacceptable effects of the market economy."[23] The architects of the self-management economy originally envisaged financial assistance being provided only between independent companies that would enter into a "self-management agreement" with one another. The 1980s, by contrast, symbolize a nationalization of solidarity in Yugoslavia; this was no longer negotiated between the so-called "organizations of associated labour" but at the republic and the national level. As a result, the state became the "ultimate risk-taker."[24] But who exactly was "the state" in Yugoslavia?

Representatives of the Croatian government attempted to convince the Yugoslav government of the need to save the shipbuilding industry. The 1974 constitution had already envisaged state actors playing a significant coordinating role between self-managed entities. The framing of the sector as national was accompanied by a new dimension of state intervention. As part of the austerity requirements of the IMF, from which Yugoslavia requested support in 1982 in the face of imminent insolvency, the system of the *visible hand* enjoyed something of a comeback in key sectors.[25]

The government restructuring loans that managed to stave of Uljanik's bankruptcy at the beginning of the 1980s were linked with an expansion of the state domain. The Law on Reorganization and Liquidation of Commercial Organizations passed by the Yugoslav parliament in 1980 required unprofitable basic organizations to present a restructuring plan to the governments of their respective republics and have the Public Accounting Office (Služba društvenog knjigovodstva, similar to an auditor) conduct an inspection to assess compliance with the plan once a quarter.[26] The introduction of this kind of reporting system in late socialism laid the foundation for a scientization of state industrial policy. Although the required restructuring measures and the related inspections tended to be relatively relaxed, they still represented a new relationship of dependence between businesses and the state. The moment the financial stability of a work organization was no longer guaranteed, state actors were automatically granted considerable rights of scrutiny and were permitted to infringe on the autonomy of the company or the very self-managed units Yugoslav politicians had been so proud of for decades. Uljanik's director, Karlo Radolović, succinctly summed up the predicament his shipyard also found itself in: "If there is no money, there is no autonomy."[27]

The democratically elected governments that came to power in Poland and Croatia in 1989/90 faced a similar dilemma to that of the communist reformers. They initially attempted to smooth the way for the shipyards to enter the market economy by means of a massive debt reduction. The dialectic between austerity and protectionism thus certainly did not end when the two shipyards became private limited companies. Even after the political revolution, the boundaries between the shipyard and the state remained porous. In 1995, the Croatian government launched a costly restructuring program. By establishing boards of creditors, the government at the time effectively took direct control of the shipyards – as the Rijeka daily *Novi list* aptly reported, the state went from "being an owner on paper to being an owner in reality,"[28] although it did not act like one. In 2000, the state troubleshooter once again offloaded the debts that Uljanik had accrued with the private banks onto the Government Agency for Bank Recovery (Državna agencija za sanaciju banaka).[29] At the same time, the government exempted the shipyard from its employer social security contributions and waived the overdue loan interest payments owing to the Ministry of Finance as well as the Ministry of the Economy's receivables. The mechanism of collective liability proliferated more and more and began to have a negative impact on more resilient companies in other sectors, such as the postal service. By 2012, a total of almost 30 billion kuna (four billion euros) in state aid had gone into shipbuilding in independent Croatia. On the eve of EU accession, the rules of the game fundamentally changed. But even after accession, the Croatian state still bore the risk, with the government continuing to grant guarantees for the loans that the shipyards needed to finance the construction of new vessels.

The socialization of corporate debt in post-socialist Croatia resembled the mechanisms used to distribute Gdynia Shipyard's debts throughout Polish society after the upheaval of 1989, the most frequently seen method being "bad banking" – in the form of offloading corporate debt onto state banks, especially the Polish national bank. The "bad bank" mechanism originated in socialist practices. In the 1980s, the Ministry for the Steel and Machine Construction Industry effected a loan from the national bank to the Paris Commune Shipyard. Despite the fact that the shipyard's liabilities to this very same bank were already increasing by four million złoty per day and everyone involved was aware that the shipyard would not be in a position to pay back the loan any time soon, state actors urged the bank to provide the shipyard with liquidity.[30] As well as the banks, public utility companies and social security organizations also bore the burden of the ailing company. In 1995, for example, Gdynia Shipyard owed ZUS 111 billion złoty, making it the institution's biggest debtor in the entire province of Gdańsk. In 2004, too, the government restructured part of the debt resulting from outstanding contributions to ZUS and the insurance company PZU.[31] An even more pressing problem in the early 1990s was unpaid energy bills. And while the state-owned energy company Energa Group did power down the electricity supply, it still declared its solidarity with the shipyard and categorically ruled out ever cutting off its power supply completely.[32] In Pula, too, despite symbolically disconnecting the shipyard every once in a while, the municipal utility company never followed through with its threat to completely take Uljanik off the grid because of its arrears.

Thus, the boundaries between the shipyard and the different levels of government remained blurred. Even after the change in political system, continuities can be identified between late socialism and the tentative decision-making of the first reform phase. Every attempt at privatization came up against a countermovement "from below," which reinforced the state's position as the protector. In this respect, the social and political relations of power limited the reformers' wish for structural change – that is, until EU accession. The consistently high degree of entanglement manifested itself in the shipyards' special political status: they were not hermetically sealed industrial organizations but rather objects of public interest. And they acted accordingly by quite readily giving the public insight into their business in order to encourage a local spirit of solidarity. It was precisely this special position that made the privatization and subsequent closure of the two shipyards so difficult. The unclear boundary lines made the consequences of insolvency, both socially and spatially, difficult to predict.

In times of crisis, the myriad interrelations between company and state took on a life of their own anyway: control became political responsibility and liability became compensation. With each loss-making year and each bailout package, the noose around the necks of the members of government who shared the responsibility became tighter, until they found themselves well and truly stuck between a rock and a hard place. The state-subsidized restructuring and ultimately liquidation of the shipyards were both extremely expensive. The main difference between the two options was in their temporality: how far did the government want to kick the can further down the road, shifting the social costs of such a decision to the future – in other words, making the future generations of taxpayers foot the bill? Here, at least in retrospect, transformation seems like protracted structural change, with the scope for action narrowing year by year without the actors really being aware of it – until the market and the authorities in Brussels rudely awakened them from their daydream of continuing their familiar and well-loved practices. Although the specific "temporal fix" (to paraphrase David Harvey)[33] of late state socialism, which simply involved transferring liabilities to the future without creating assets at the same time, had continued to some degree beyond the end of communism, it was finally reaching its limits. And this remained true despite the fact that Western creditors were prepared to write off the old debt accrued during the communist era for the new democratic governments, something that Poland, in particular, profited from.

In light of these interdependencies at the intersection of control and liability, is the concept of the "company" – as an embedded but fundamentally still independent organizational unit – even still fitting? For long periods, the web of relationships between the state and business was so watertight that a definition of companies as areas of relatively autonomous decision-making and self-contained accounting was indeed unsuitable for the two shipyards. In order to elucidate this hybrid position, in the 1980s, Polish sociologist Tomasz Żukowski coined the phrase "factory bureaucracy."[34] This understanding of the company has lost none of its relevance, even in an era of neoliberal hegemony in which firms are seen as free-floating stock-market entities that are responsible for themselves. The example of the shipyards' financial dependence discussed here showed that key company decisions were debated within a complex and dynamic group of stakeholders comprising state organizations, other social actors, as well as the company's interest groups.

The Meandering Leviathan

> All our workers have heard about Jadranbrod's transformation [*transformacija*]; everyone knows that their basic organization of associated labour belongs to one of the groups of this great Yugoslav shipbuilding community. They all attended a workers' conference at which they were familiarized with the reasons for such an association and were given the opportunity to air their views on it.[35]

As this article in the trade journal *Brodograditelj* from 1976 shows, the period of political upheaval after 1989 by no means had the monopoly on the concept of transformation, at least not in Yugoslavia. When the trade press reported on "transformation" in the 1970s, they were generally referring to the industry and company level.[36] The concept, which stemmed from structuralist vocabulary, was used to describe a qualitative change that altered the very essence of the industry association Jadranbrod. Underpinning this change was the break-up of companies facilitated by the 1974 constitution and the 1976 Law on Associated Labour. This atomization made stronger coordination at the sectoral level essential. It also provided an alternative to the failed attempt to integrate Yugoslavia's shipyards into a conglomerate, a move that would have created new opportunities for the division of labour and specialization but one the workers of Uljanik voted against in 1967.[37]

Jadranbrod, which had grown out of the Udruženje brodograđevne industrije (est. 1956), was situated in a sphere of activity that was somewhere between the shipyards and state actors. Political scientists' discovery of the heterogeneity of the state sector in the 1990s was all the more applicable to the centrifugal Yugoslavian system.[38] Driven by political reforms and the struggle for resources between companies and governments, the makeup of the different actors changed for good. New actors appeared, established actors took on new and unfamiliar functions, while others lost theirs altogether. After the first phase of the *transformacija* in the 1970s, Jadranbrod comprised twenty-four work organizations with a total of around twenty-five thousand employees.[39] The idea behind an industry association was for it to fulfil or at least coordinate any and all tasks that very small basic organizations found difficult to manage on their own, such as research and development, international marketing, comparative analyses, and training.[40] Transferring these activities to the industry level served to counter the excesses of market socialism and rampant plant egoism. Political actors took control, while directors' and companies' room for manoeuvre was severely restricted.

In 1976, Jadranbrod had to adapt its organizational form once again – this time to the Law on Associated Labour and the new economic reality following the downturn of the global shipping market. Economic experts saw the fragmentation of the basic organizations as the reason why Yugoslavia's shipyards were unable to cope with the new market situation. They argued that investments were obstructed by the boundary drawn between the different business units. In response, Jadranbrod attempted to create more coherence again: the industry association's members should not be the individual basic organizations but the shipyards in their entirety. This should – contrary to the pluralist approach of self-management – recreate a vertical structure, with Jadanbrod at the head. Unlike in the Polish People's Republic, the Yugoslav industry association was, however, responsible only for coordination and setting cross-company standards but unequivocally not for planning activities.[41]

However, this increase in coordination and oversight at the sectoral level, while at the same time preserving self-management at the company level, resulted only in even more incoherence and contradictions, as local actors never tired of pointing out – and which the company newspaper repeatedly lampooned (see figure 3.1). Moreover, Jadranbrod lacked any real power to enforce anything. In an interview in 1986, Uljanik's director criticized the lack of accountability that had established itself in shipbuilding. On the one hand, he was concerned about the weak position of the so-called complex organization of associated labour (the term used to describe the company as a whole) compared with the smaller basic work organizations, which were nevertheless fundamental as defined by law. On the other hand, he criticized Jadranbrod for being a "rigid system" that afforded the shipyards less and less room for manoeuvre.[42] More than anything, however, Uljanik's director bemoaned the ambivalent role of the state in the 1980s. Although it had an increasingly strong supervisory function, because of the IMF conditions, the state was now prepared to take responsibility and provide fresh capital only to a limited extent:

> I then told the President of the Executive Board that I am totally in favour [of large, rigid systems], but responsibility for paying the wages has to lie with Zagreb. And if people aren't paid and decide to go on strike, they should go to Zagreb and not come to me.[43]

Figure 3.1. *Uncertainty to the Very End*, from Uljanik company newspaper, 1982
Source: Printed in *Uljanik*, no. 13, 1982. Courtesy of Historical and Maritime Museum of Istria.

The state actors involved in Poland's Paris Commune Shipyard endured a similarly ambivalent transformation. The economic reforms of 1981 pulled the rug out from under the feet of the line organization, from the planning committee to the factory floor. In the typical model of the command economy, the industry association was responsible for transmitting guidelines between the ministry and the company. It fell to them "to further flesh out, negotiate, and implement the requirements of the plan in both directions," states a recent account.[44] In the first blueprint, inspired by Solidarność, the industry associations were to be removed entirely from hierarchical economic planning. This radical reform model soon became cause for concern, especially in the shipbuilding industry. For the head of the Institute of Ocean Engineering and Ship Technology at Gdańsk University of Technology, Jerzy Doerffer, it was an illusion to think that the companies in the shipbuilding industry could be fully self-sufficient:

> Should a particular shipyard refrain from becoming a member of a federation or an association, they would automatically be cut off from information and the protection of powerful associations, and, in today's world, this is synonymous with bidding farewell to technological excellence.[45]

A second problem was the value chains between the individual companies. With the loss of industry coordination, the shipyard faced protracted negotiations with suppliers like Poznan-based turbine manufacturers Cegielski. Nevertheless, in 1981, the Polish Council of Ministers decided to abolish the industry associations, a decision that also meant the end of the professional organization of the shipbuilding industry (Zjednoczenie przemysłu Okrętowego). In their place, following the imposition of martial law, the government established the Association of Companies in the Shipbuilding Industry (Zrzeszenie przedsiębiorstw przemysłu okrętowego), membership of which was compulsory for a transitional period of five years. The director of this association was keen to emphasize the purportedly democratic basis and supportive nature of his organization, claiming the association's "positive intentions and a willingness to cooperate."[46] Much like its predecessor, however, the association continued to have considerable influence over the distribution of the limited supply of raw materials, goods, and foreign currency, and was thus an indispensable authority. According to a historical evaluation of the reforms, "The process of liquidation of the industry associations, which was to initiate the de-monopolization and restructuring of the industry, thus remained fictitious."[47]

The five-year mandatory membership was not enough to ensure long-term stability for this fragile industry body. When, in 1987, more and more companies began to leave the association, the director, in an attempt to save it, proposed various reforms, perhaps the most radical being to restructure it to create a corporate group.[48] At a meeting of the association council in June 1988, besides this proposal, there were various other organizational concepts on the table: continuing as an association, a centrally controlled combine, a collective akin to the industry association that had once existed, a corporation, and a holding, the latter being an organizational form that was unprecedented in the Polish economy, however.[49] Even Solidarność, which had in the past always insisted on the independence of the companies now began to doubt the practicality of its demand.

In 1989, the companies were unable to reach a consensus on the reorganization of the industry, leaving them with an organizational and institutional vacuum.[50] Describing this period, the aforementioned Jerzy Doerffer, who went on to found the employers' association the Association of Polish Maritime Industries Forum Okrętowe, wrote, "The complete absence of any kind of coordination between the individual shipyards resulted in total chaos."[51] To make sure the companies' concerns did not completely fall by the wayside, a group of shipbuilding experts formed an advisory board to support the deputy minister for transport and the shipbuilding industry in the first democratically elected government. In 1991, the directors ultimately created their own "Forum." At long last the requirements of the economic reforms implemented in 1980/1 seemed to have been fulfilled. Membership of the Forum was voluntary, no longer based on state-controlled ownership, and had shrugged off all forms of supervision. On the one hand, the organizational form of an employers' association was a good fit for the new neoliberal environment. The association provided expertise and knowledge in order to assist the restructuring of the industry as well as to support the individual companies, and, with its political contacts, it also formed a powerful lobby. The post-socialist path dependencies were further strengthened with the creation of the social partnership provided for in the "Pact on State-Owned Enterprises Undergoing Restructuring" concluded in 1993, paving the way for protection mechanisms negotiated on the basis of a corporatist model in the shipbuilding industry, too.[52]

In the late 1980s, Croatia's Jadranbrod developed in a similar direction when, with the individual shipyards no longer willing to finance the industry association, it looked as though it would gently meet its end. Symbolic of the crisis in the association was the fact that in 1989, it published only a single edition of its journal *Brodogradnja*, and none at all in 1990. In 1994, the Croatian government created the association

Hrvatska brodogradnja (Croatian Shipbuilding, generally abbreviated to "HB") in the form of a registered company, and in 1997 this was merged with Jadranbrod. From this point on, the once self-managed industry association no longer represented the shipyards, acting instead as an extended arm of the government, which utilized HB's expertise and entrusted it with the oversight of the state-owned shipyards. Following the privatization of the shipyards over the course of the EU accession process, the government tasked the organization with evaluating the shipyards' applications for loan guarantees. Since 2013, HB has officially been on the list of companies of "strategic and particular interest for the Republic of Croatia."[53]

With the end of communist rule, forms of ownership changed, as did the actors responsible for implementing the current political and economic goals in the shipyards, which were still (or once again) in the hands of the state. Both in Poland and Croatia, the 1990s saw the creation of a whole series of new economic policy organizations that would go on to help shape the fate of the shipyards. After Croatian independence, the Tuđman government continued along the path of nationalization taken by Ante Marković, the last prime minister of Yugoslavia. The governing centre-right Croatian Democratic Union (HDZ) passed a legislative package providing for the establishment of the Agency for Restructuring and Development. The aim of the agency was to safeguard "the national interests of the Republic of Croatia" during the transfer of ownership.[54] It was this agency that was to decide on Uljanik's application to re-establish itself as a joint-stock company. The ownership rights went to what was known as the Croatian Development Fund before being transferred to the newly created Croatian Privatization Fund at the end of 1992. The problem with these funds was that they were conceived as silent transitional owners and as such had no development strategy for the company.[55] In 1990, however, the government did meet Karlo Radolović's request to establish a Ministry for Maritime Affairs. That said, when interviewed, the new minister, a non-partisan professor of international law and close aide of Tuđman, emphasized that "the government does not want to become a shipyard owner"[56] – and yet the shipyard was actually already state-owned property. The result of this was that ownership policy and industrial policy were in the hands of two different organizations and only loosely connected.

The government repeatedly announced that it was going to privatize Uljanik, but failed to systematically address the issue. Instead, the shipyard remained tied to the state through a variety of public co-owners, who had become shareholders after the shipyard's debts had been converted to shares. The largest of these owners remained the Croatian Privatization Fund, which at the start of the 2000s had received 36 per cent of the company's shares; second in line was the state Agency for Bank Rehabilitation (27 per cent), followed by the state pension fund (13.6 per cent), and state health insurance fund (7 per cent). There was certainly no shortage of willingness on the part of the state owner to inject capital into the shipyard. What was far more critical, however, was the parallel failure of the owners to exercise oversight. At no point was it clear from the portfolio of the Croatian Privatization Fund which companies were earmarked for privatization and which were, for "strategic reasons," to remain under state control. On top of this, the decisions made by the state owners often lacked transparency and the oversight provided by the supervisory boards was frequently unprofessional.[57]

With Croatia's advancing integration into supranational economic contexts – it joined the World Trade Organization (WTO) in 2000 and signed the Stability and Association Agreement with the EU and began accession talks in 2003 – the bloated state industrial sector and the conflation of industrial and social policy increasingly became bones of contention. Even before EU accession, these frictions made their influence felt on the government stock portfolio. In 2010, the HDZ government replaced the Privatization Fund with the Agency for State Property Management and began to thin out and consolidate its portfolio further.[58] One of the consequences of this was that Uljanik was removed from the government stock portfolio – in 2012/13, Uljanik was privatized by making the employees shareholders. For the EU, which had learned its lesson from Gdynia, this privatization was a strict requirement for Croatia's accession to the European Union.

The case of Croatia confirms the conclusion drawn by two prominent transformation researchers that the problem in the post-socialist countries of the 1990s was less related to the legacy of a strong state and more the result of that state's weakness.[59] In her comparative analysis of the restructuring of steelworks in Eastern Europe, political scientist Aleksandra Sznajder Lee identified "state capacity" as a key factor determining the success or failure of corporate reforms. Only efficient state institutions deemed legitimate by those affected were able to systematically push through and implement the changes in erstwhile state companies that were needed for them to adapt to a private market economy.[60] Political scientist Grzegorz Ekiert argues along similar lines: "Communism left behind not a powerful bureaucratic Leviathan but a weak and inefficient state."[61]

In Poland, too, this observation applies not only to the political sphere but also to state-owned industry, where, after 1989, the state emerged as an owner with no vision, no expertise, and ultimately no capital. The government's restraint was rooted primarily in something akin to ideological self-limitation.[62] Owing to the one-dimensional focus on privatization, the owners were not really interested in influencing microeconomic processes or developing and implementing longer-term strategies for the shipyard. The actions of the Ministry for Ownership Transformation (Ministerstwo Przekształceń Własnościowych) were more on a par with that of a non-owner. As economist Andrzej Kensbok has pointed out, the void where the owner should have been had fatal consequences for the company: "They [the companies] often lack the strategic vision, room for manoeuvre, and leadership that would arise out of a sense of ownership."[63] In this sense, the neoliberal credo of the inherent inefficiency of state-owned enterprises became something of a self-fulfilling prophecy. Only in the area of debt management did the state prove to be capable of action.[64] The government increasingly geared its actions towards the company's and workers' need for protection, while the reforms that were originally planned fell by the wayside.

In 1996, the government headed by the post-communist Democratic Left Alliance (Sojusz Lewicy Demokratycznej, SLD) used the dilemma of how to deal with state-owned joint-stock companies as an opportunity to implement what was known as the "centre reform." The government replaced the Ministry for Ownership Transformation with the Ministry of State Treasury (Ministerstwo Skarbu Państwa). Thus, the reform sought less to discontinue subsidies but rather to strengthen state control and management of subsidized companies. Unlike the Ministry for Ownership Transformation, which had only played the role of assessor in the process of privatization, the new ministry integrated the "management of state assets" into its portfolio, seeking to "defend the interests of the state."[65]

The downside of this attempt at reinstating owner control and oversight was the politicization of companies. Each parliamentary election meant another turn of the "carousel" of ministerial posts, with board members being replaced and CEOs showing clear political affiliations. From 2004, Gdynia Shipyard felt the fatal fallout of this process of politicization when the state once again became the company's biggest shareholder. This was in striking contrast to the situation in Pula, where, despite the state effectively being the majority shareholder up until privatization, from 2012, there was little evidence of direct government influence on the board. Although long-time director Radolović did become a member of the governing HDZ, beyond the clientelism that was typical of Croatian capitalism, party politics does not appear to have played a role. This separation of spheres of action is an indication of the path dependency of Yugoslav self-management.

With the state as majority shareholder – the government refused to call it (re)nationalization – from 2004, the shipyard in Gdynia essentially gained easier access to state aid. In the very same year, Poland became a member of the EU and from then on had to submit to the regulations and directives of European competition law (see below). The European Commission required Poland not only to reduce its protectionist measures but also to strengthen its ownership policy. The "White Book" by SLD minister of state treasury Wiesław Kaczmarek took a similar line, denouncing companies' shortcomings and calling for increased oversight. Yet this still did not answer the fundamental question of what the objectives of the state owners were. The predominant position was still one of defensive legitimation, which justified state-owned property as a defence mechanism against the unwelcome effects of the free market. What was needed here was outside intervention from the Commission – in other words, for it to demand clearly defined objectives. The shipyard in Gdynia was thus not simply a "sacrificial pawn" in the process of EU accession but also "collateral damage" of the Polish state's failure to advance and modernize the country's business policy. This outcome was not necessarily a given, as seen in the fate of the Polish steel industry, whose privatization was a condition of EU accession. In this case, thanks to wealthy foreign investors, as well as a successful social dialogue, it had been possible to more or less maintain production levels despite the almost one-third reduction in the workforce.[66]

In 2009, one year after Gdynia Shipyard went into liquidation, the government published a very selective share portfolio of the Polish state under the title "National Program for Owner Control."[67] Shipbuilding was no longer seen as a key future industry that the government promised to protect from free-market forces. The "good change" that the government of the right-wing populist Law and Justice Party (PiS) hailed on taking office in 2015 was to turn back the wheels of time and put the EU competition authorities back in their place. During its election campaign, PiS promised to establish a program of state funding for the shipyards, to reopen the shipyards in Gdynia and Szczecin – and all of this would ideally be financed by the European Union.[68]

This analysis of the actions taken by the state contradicts the widely held view of a linear transformation from a command to a market economy. The dialectic of reform initiatives and protectionist reactions is clearly illustrated by our two case studies. We do not consider these two cases to be exceptions but rather see the meandering course of politics and business as a typical path of development that becomes evident only when we analyse a longer period and look specifically at the company level. All of this reminds us of the pendular movement and the counterreaction against the unfettered market that Karl Polanyi describes in his book *The Great Transformation*. Patchy reforms, a lack of state capacity, opaque ownership and industrial policy, and a lack of commitment to EU competition policy resulting from social and political considerations – these are all typical characteristics of post-socialist transformation in Eastern Europe. In this field full of grey areas, the actors involved had to come up with excellent reasoning for their interests. Here, too, the deeply ambivalent character of transformation comes to light; its direction was negotiated, the reform vector constantly realigned in the face of recurrent resistance, while the term "structural change" evokes the image of a certain inevitability the instant the lava of change begins to flow.

Rhetorics of Legitimation and Mechanisms of Social Protection

"Requiescat in pace." This was the inscription on the black coffin that trade unionists carried through Warsaw to the office of the Polish prime minister on 12 May 2008. The protest march was the last in a long series of visits to the capital organized by workers from the shipyard on the Baltic. As in August 1980, the trade unions initiated a round of talks between the employees and the government shortly before the company was liquidated. Once again, the workers went over the head of their employer, who in this negotiation scenario was reduced to a "sideline player."[69] Unlike the trade unionists in Pula, their Polish counterparts were used to the way negotiations were handled under state socialism. Accordingly, their behaviour immediately showed politicization on two levels: first, the party-political bias of the trade unions, as pointed out by political scientist David Ost, sent Solidarność on a collision course with the post-communist SLD government.[70] Second, and to date very little research has been done on this, was the tendency to shift conflicts to a level where the state manages industrial relations and conflicts.

The direct communication between the workforce and the government, where the middleman is cut out, followed a pattern that had also been typical for the pre-1989 period. Hungarian sociologist Teréz Laky identified what he refers to as "plan bargaining" in the context of socialist economies – as opposed to the "shopfloor bargaining" between trade union representatives and management boards at the company level as seen in the market economy. Laky defined this "plan bargaining" as negotiations conducted between companies and the state administration centring on the individual plan goals, the allocation of state resources, and after the first oil price shock, aid programs. According to Laky, the outcome of these negotiations depends on the structural, economic, and instrumental power resources of the participating actors. But what narratives did the company and its various stakeholders employ during these negotiations in order to legitimize their demands? It is this question that the next section seeks to answer.

In the Polish People's Republic, under the banner of economic decentralization led by party head Edward Gierek, the socialist combines had continued to grow unabated. They had been able to consolidate their monopoly-based structural power, especially since the industry ministries represented their interests in the government.[71] Although the Paris Commune yard in Gdynia was just one of six Polish shipyards, its portfolio and production capacity were unparalleled. A second dry dock and the corresponding technical infrastructure made the Paris Commune yard the most modern shipbuilder in Poland, enabling it to specialize in the construction of bulk carriers and tankers. Its negotiating power was rooted in its "technological monopoly,"[72] as the foreign trade company Centromor relied on Gdynia Shipyard's production equipment to complete quite a number of its foreign orders.

Unlike in capitalist countries, the companies under state socialism had a genuine interest in permanently increasing their workforce. The majority of arguments presented in the political economy literature on "labour hoarding" under socialism refer either to the inefficient socialist companies' insatiable appetite for labour in an economy with full employment or to the need to provide a buffer against external imponderables such as the requirements of five-year plans or sudden deliveries of long-awaited upstream supplies.[73] For cases like this, which were far from a rare occurrence, reserves of workers were absolutely essential. These two phenomena are described by sociologists Charles Sabel and David Stark as "defensive" strategies.[74] Conversely, the socialist companies could also use employment as an offensive strategy and power resource: "Command over resources is a sign of political status in these economies, and this applies especially to command over labour."[75] The more employees a company had, the more compelling its argument that it was essential for local social stability and the formation of a socialist working class – the persuasiveness of this argument increased exponentially under socialism owing to the numerous welfare services the companies provided for the local communities.[76]

During Gierek's decade in office from 1970 to 1980 (see figure 3.2), in line with this logic, the Paris Commune Shipyard's workforce grew to well over 12,000. In the 1970s, it was Poland's second-largest shipyard after the Lenin shipyard in Gdańsk, placing it on an equal footing with the Warski shipyard in Szczecin. In 1982, all shipyards

Figure 3.2. Two workers reading a newspaper report about Edward Gierek's appointment as first secretary of the Polish United Workers' Party, 1970
Source: Courtesy of European Solidarity Centre, © Janusz Uklejewski.

combined were the maritime industry's biggest employers, between them providing jobs for 39,000 people.[77] Another 17,000 people worked for direct suppliers, which were also members of the shipbuilding association.[78] The Yugoslav shipyards pursued a very similar strategy: despite the slump in sales in the 1980s, their workforce continued to grow (in the case of Uljanik to over 8,000 workers) and with it their socio-political capital in negotiations on state aid.

The second dimension of the shipyards' distinctive economic power resources was their importance for the foreign trade balance. In Poland, for instance, since the 1970s, Gdynia Shipyard had become increasingly invaluable for generating the foreign currency the country so urgently needed, as the international demand for Polish coal stagnated and this sector's share of the country's foreign currency revenues fell to 20 per cent. At the same time, external debt resulting, for instance, from technology imported from the West, grew exponentially, and with it foreign debt service payments in US dollars. In response to this situation, the government went in search of new export goods. Despite the industry's declining shares of foreign trade overall, Gdynia remained an important exporter well beyond the 1980s, distinguishing themselves from the other shipyards. In the 1986 Lloyd's Register, the company occupied the top position among the Polish shipyards when it came to seaborne exports and ranked tenth among all Polish companies as regards export value.[79] Investing in a modern dry dock seemed to have paid off.

Uljanik, too, availed itself of the export mantra to justify its demanding attitude. Its management never tired of referring to the shipyard's role as an earner of foreign currency. In an interview given in 1986, director Karlo Radolović pointed out that with its around 200 million US dollars in annual revenues, Uljanik was one of Yugoslavia's most successful exporting companies (he neglected to provide any information on the cost side of things).[80] By the mid-1980s, the shipyard ranked tenth in the list of Yugoslav exporters.[81] If nothing else, the local public and politicians found the reference to this strength to be an extremely convincing justification for the level of state care and attention the shipyard was receiving. The local newspaper *Glas Istre* was full of articles depicting Uljanik's export successes; for the city of Pula and Uljanik's workers, these were a source of great pride.

The third traditional power resource the company had at its disposal, instrumental power, relied heavily on individuals – in Uljanik's case, long-time director Karlo Radolović. Given that in socialist economies, prices played only a subordinate role in economic coordination, informal processes of negotiation, and thus also networks, as well as the charisma of individual personalities were particularly important. A lot depended on the skills of these individuals, although the structures of power were just as significant. Nurturing and maintaining political relationships was one of the management's absolute priorities – both during and after socialism. According to historian Maciej Tymiński, the negotiating power of company directors in a command economy very much came down to what the individual directors brought to the table: their specific political connections as well as the support of local power structures and the factory organizations.[82] Shop floor managers focused their attention on negotiating the political field, not on the company's balance sheet.[83]

In Gdynia, the shipyard's instrumental negotiating power relied on Zbigniew Maciejewski. The once deputy technical director and later deputy production director, Maciejewski took the helm in Gdynia at the end of 1981 after the Ministry for the Steel and Machine Construction Industry deposed his predecessor in the process of introducing martial law in December of that year. There is ample anecdotal evidence that long-time workers credited Maciejewski with creative problem-solving skills. One example of this aptitude was the purchase of fifty Sinclair ZX Spectrum computers that a shipowner insisted on having on their vessel. The problem was that these computers were on what was called the CoCom list, which served to limit the export of dual-use Western technology to Eastern Bloc countries. Nevertheless, the director promised to fulfil the request, saying, "I'll handle the customs authorities." Not only did the computers somehow get through British customs; Maciejewski also decided to get some human capital out of his coup. Before installing the new machines, his best workers were each given a computer (off the books) to take home for a month to practice – now that is what you call staff training in an economy of scarcity!

Even in Yugoslavia with its system of self-management, in which the main arena for industrial negotiations was essentially the company itself, the directors still played a key role. They embodied the relationship with the state bureaucracy that was so crucial for accessing state resources (such as foreign loans or the special drawing rights offered by the central bank).[84] Successful directors are distinguished by their ability to form alliances – both vertical and horizontal. Although the 1974 constitution and the 1976 Law on Associated Labour meant the so-called "red" directors lost their true responsibility and autonomy when it came to staffing, well-connected and well-respected managers such as Radolović were in a position to play a role similar to that of Western CEOs.[85] In their analysis of the function of the director in socialist Yugoslavia, Jurij Fikfak and his co-authors describe directors as "some sort of external affairs ministers of their own workers' councils, a link between broader social interests and the interests of the working organization."[86] This role demanded a high level of political skill as well as recognition and respect within the company itself. In this sense, Radolović was a shining example of a Yugoslav director and his skills were put to good use, even after 1990. Much like the shipyard in Gdynia, Uljanik, too, managed to get the state actors to open their coffers in order to plug the holes in the company's finances; but it was not enough to guarantee anything more than the shipyards' survival. What Kornai concluded in the abstract in his analysis of the repercussions of "soft budget constraints" for business activity thus also proved to be true in economic life. Factory managers often invested more energy in nurturing political relations than increasing productivity, as their company's success depended more on the goodwill of the state than on their profit margin.

In Poland, the shipyards' exceptional negotiating position already aroused some envy in the 1980s. The "lobby on the slipway"[87] was now increasingly the target of criticism by reformers or representatives of other sectors. They called for the government to stop mollycoddling the shipyards and for them to be exposed to the market. Thus it seemed that the convertibility of the traditional power resources into financial aid was not without its limits. In Yugoslavia, too, the lobbyists for the shipbuilding industry were well aware of the problematic nature of the seemingly objective export earnings – as shown above, Uljanik rarely made a profit from the ships it produced for foreign customers in the 1980s. Politically symbolic resources, in contrast, became more and more important. Indeed, irrespective of its economic success, the people working at Uljanik only confirmed the global allure of their shipyard. In an interview in 1992, Karlo Radolović joyfully announced that he had managed to teach President Tuđman as well as the prime minister what "Croatian shipbuilding and its world trademark actually mean."[88] After the end of communism, the shipyards' stakeholders worked ever more intensively on producing narratives to highlight the significance of their industry for the country and the region.

Given that the 1990s saw no improvement in the shipyards' economic situation, the pressure on the industry to legitimize its existence remained high. The fact that in Croatia's case, the state had, in 1991, become a shipyard owner for the first time only increased this pressure. After Uljanik was nationalized, employees, trade unionists, and the executive board used a veritable "polyphony"[89] of arguments, depending on the time and context, to improve their negotiating position towards the Croatian government, with themes ranging from identity politics to day-to-day political issues. In the early 1990s, for instance, the shipyards jumped on the nationalist bandwagon of the Tuđman government, duly playing up their war efforts. According to Milan Čuvalo, an industry expert and adviser to the minister for economic affairs, shipbuilding companies had proven that they were not afraid of defending their homeland: "When it became necessary to go to the first line of defence, to build a massive bridge, or to cut wages, the shipbuilders were the first there."[90]

Radolović highlighted the organic link between the shipyard and the nation, underlining that Uljanik represented the entire state *pars pro toto*. The shipyard could not exist without a healthy and peaceful state because it depended on a steady inflow of orders as well as on foreign loans. And the state, in turn, could not flourish without the shipyard because the latter represented social welfare and stability. Radolović asserted that he had informed the government leadership in 1992 that shipbuilding could make an important contribution to achieving full employment after the war. In fact, shipbuilding was the only industry with any prospect of orders and would thus be able to provide jobs for demobilized soldiers on their return home, he explained.[91] Consequently, according to an article in *Glas Istre* that equated the prosperity of the shipbuilding industry with that of the state, the government has to "create the conditions for the state, and thus also the shipbuilding industry, to be able to function."[92] Radolović managed to infect the shareholders with his rhetoric, declaring at Uljanik's 1994 general assembly, "Without state aid, the shipbuilding industry will not be able to surmount its difficulties, but nor will the Croatian state be able to resolve its economic problems if the industry doesn't recover."[93]

The patriotic line, however, was not the only one taken by the shipbuilders to justify their demands. Another argument was directed at those in society who were not susceptible to a nationalist ideology – a relatively large group in Istria, a region that is proud of its multiculturalism as well as its distinctive regional consciousness. Since Croatian independence, Istria has been governed by an explicitly anti-nationalist, social democratic leaning regionalist party, the IDS. Consequently, at the local level, it was more socio-political narratives that took hold.[94] For one representative of an Istrian trade union, it was apparent "that a Croatia that is only greedy for profit, without protecting jobs and the workers to the same extent, would not be a true state."[95] A legitimate state would have to make a break with the IMF's conditions and the shock therapy approach pursued by the last prime minster of Yugoslavia, Ante Marković, in order to keep the number of losers in the process of transformation to a minimum. Many trade unionists regarded social policy as a litmus test for the government's effectiveness. Representatives of this position called for social policy aspects to be taken into account during the restructuring of the shipyard and for a large workforce to be maintained. This corresponds to the idea of "strategic social policy," which attempts to resolve certain issues before they even arise in the first place, an approach applied by Danish political scientist Pieter Vanhuysse to the context of transformation. After all, during the war, the Croatian government did not want a fight on the social front as well.[96]

Further, in the above-mentioned interview, Radolović also mentioned the idea of using trade policy – future orders and new export successes –

as a resource for legitimation towards the newly independent, war-ravaged, and virtually bankrupt state. The director admitted that the current order situation was disastrous, and the shipyard was essentially insolvent. But was it not the war that was mainly to blame for this? The world economy, and with it the demand for ships, were on an upward trend, and Radolović thus presented his shipyard as a promise to the Croatian state about the future. In a similar vein, the industry adviser to the Minster of Economic Affairs never tired of emphasizing the fantastic prospects of shipbuilding. He anticipated

> that the following year [1994] would be better in every single respect. Slowly but surely, the political and economic conditions in Croatia are becoming more stable, shipbuilding is expecting a boom.... Along with orders, jobs, and savings, this is cause for optimism and hope for a better tomorrow.[97]

The polyphony of arguments continued to prove effective in helping the company to avoid insolvency when it found itself in serious crisis in the mid-1990s. Not only was it possible to push through state-financed restructuring but the shipyard also managed to get the state on board as a financier of last resort beyond the turn of the millennium. In 2017, despite its by that point completed privatization, Uljanik asked the state to bail it out yet again as it faced the next, and ultimately the final, major crisis. Initially at least, there was a cross-party consensus that the shipyards in Pula and Rijeka, both of which were then owned by Uljanik, had to be saved.

This decision was based on a slightly updated version of the familiar set of arguments.[98] In a statement in March 2018, Croatia's minister for economic affairs expressed her wish to keep Pula from becoming a "monotown" whose fate depends solely on tourism.[99] Other politicians declared the preservation of shipbuilding to be a matter of "national interest," without going into further detail, however.[100] And others still described it as a "strategic industry," though here, too, it remained unclear what exactly this implied.[101] Whereas local politicians emphasized the social benefits of the shipyard and the governing party shied away from publicly announcing tough cuts, the workers' self-legitimation was based on their own day-to-day work experiences. For them, full order books and a high production volume made the question of profit obsolete. The general thrust of their argument was "If Uljanik uses its resources well, but still makes a loss, then it deserves to receive the necessary state support to survive."[102] The new CEO Gianni Rossanda highlighted that Uljanik had actually benefitted from far less state aid than all the other shipyards in the last twenty-five years, also pointing out that the reorientation of the production portfolio had already been initiated.[103] Of course, there was one possible raison d'être that never featured in this polyphony – nothing was ever said about the profit generated.

When Debt Management Becomes Blame Management

There were a great many plausible reasons to save the Croatian shipyards. Yet, in 2018, the pro-European HDZ government was slow to act, almost as though it wanted to first ascertain the shipbuilders' reaction and get a feel for the social mood. The precedent of the Polish shipyard in Gdynia in 2009 had also called the legality of potential state aid into question, particularly as the Croatian government was committed to a path of budget consolidation. In their view, this time the protection had to come from the market itself: a "strategic investor" should, like a *deus ex machina*, solve the problem of both the legitimacy and the legality of Uljanik's bailout in one fell swoop. This would, for the first time, relieve the state of its function as a protective barrier, also surmounting the dichotomy between laissez-faire (an ideology the government subscribed to) and protectionism (which society expected). Along with his finance minister, Croatia's prime minister Andrej Plenković, who had already been branded a Eurocrat by his critics, hid behind EU directives and made it quite clear that the most the government was prepared to provide were state guarantees: "The board bears the responsibility."[104] For his part, the potential investor, Danko Končar, emphasized at this point that he was neither a member of the management nor an owner of the shipyard and, as such, no one could hold him liable, which is why he did not intend to inject any capital.[105] Uljanik's CEO went along with this no liability angle: "I can vouch for the fact that, [in the management of Uljanik], there was no negligence."[106] What remained was a lot of worried but not personally liable actors, who all washed their hands of any responsibility.

In the wake of the controlled withdrawal of the participating actors, in Croatia, things went from debt management to blame management. As shown by the situation in Poland, too, the ambiguity of blame and unclearly defined responsibilities were not uncommon in the liquidation of post-socialist companies.[107] With EU accession, however, the lack of accountability became unacceptable. In 2018, when different stakeholders in Pula tried, one last time, to persuade the state to rescue Uljanik, the shipyard in Gdynia was already history. The toing and froing between privatization and nationalization, the ritual negotiations between the company and the government, and an ever-present need for subsidies had significantly weakened the confidence of the market. In an era of neoliberal policy prescriptions, state aid had the opposite effect: it left an impression of companies that were uncompetitive and thus incapable of surviving.

In Poland, the investigation into Polish state aid initiated by the European Commission in 2005 also caused palpable uncertainty in the industry. For the government at the time it became a priority to keep the "sunk costs" to a minimum in the event that the shipyard went bankrupt. Out of fear of ultimately being thrown to the wolves, the national-conservative PiS government sought to keep the shipyard's operations going on a low flame until the next change of government. In 2007, the buck was then passed to the liberal Civic Platform (Platforma Obywatelska, PO). While previous governments had used their resources for debt management, the two competing political camps invested their energy into the less costly blame management. Similar to the PiS government before them (and, from 2015, also after them), the PO administration devoted itself to all conceivable rhetorical efforts to make it clear that it was not them but rather the previous government that was to blame for the predicament of the shipyard.

Both for Uljanik and Gdynia, rhetorical defensive struggles replaced real protection mechanisms. In the background of this rhetoric was the fact that EU accession meant that governments now had far less leeway in their economic and fiscal policy actions. At the same time, society continued to direct its expectations of protection towards the government. And the anger felt in 2009 when the trade unionists took their pilgrimage to the office of the prime minister, complete with symbolic coffin, was correspondingly fierce; the complaint by the hairdresser in Pula referred to at the start of the chapter – that the government's behaviour was unworthy of a "father" – was rooted in a similar attitude. The last act of state involvement was thus the so-called social plan, which was intended to mitigate the consequences of the company's bankruptcy, as well the retreat of the state, for the lives of the dock workers.

Given the importance of what for Poland and Croatia were the new EU rules on the relationship between the state and the shipyards, it is worth briefly going into the origins of this. State subsidies for shipbuilding were already highly contested in the European Community in the early 1970s, because it was obvious that the member states had entered into a self-destructive competition against one another. In 1987 and 1990, the European Council had issued specific directives laying down clear and restrictive rules for the state aid that could be granted to the shipbuilding industry with a view to accelerating capacity reduction.[108] The European Commission regarded shipbuilding, much like the textile industry, as an unsustainable sector – in stark contrast to the aerospace sector for instance, where extensive state aid led to the birth of a European "champion" in the form of Airbus.[109]

The EU conveyed its competition credo to the outside world and attempted to persuade other shipbuilding nations to abolish subsidies too – in particular South Korea, which saw itself as a free-market economy but still provided its shipbuilders with a huge amount of financial support. When South Korea refused to make any concessions, in 2002, the European Council permitted member states to provide temporary state aid to their shipyards, which were suffering due to "unfair Korean competition."[110] The Council deemed subsidies of up to 6 per cent of the agreed price for container ships and chemical and gas tankers acceptable, provided the Korean shipyards were offering to fulfil the orders for less. However, the Council made it clear that this was a temporary and extraordinary measure, and the regulation expired in March 2005.

The European Union and its big shipbuilding nations had obviously resigned themselves to the fact that they could not compete on price with the three East Asian producers that dominated the market, in particular when it came to the construction of the less technically advanced ships. European shipping companies imported more freighters than were built in Europe: according to data from the OECD, between 2007 and 2017, shipyards in the EU accounted for a little over 6 per cent of worldwide production, while European companies were the buyers in 32 per cent of vessel sales.[111] In a strategy paper published in 2003 ("LeaderSHIP 2015 – Defining the Future of the European Shipbuilding and Repair Industry – Competitiveness through Excellence"),[112] the Commission pledged that in the OECD and WTO as well as bilateral negotiations with South Korea, it would make sustained efforts to counter unfair competition. That said, in the decades preceding this, a similar undertaking with regards to Japan had come to nothing.

At the same time, the Commission recommended measures to increase competitiveness, a strategy that goes back to the European Council's conclusions from 2001 and the 2000 Lisbon Agenda, in which the EU proposed a neoliberal reform agenda. The origins of the focus on "competitiveness" and the increasingly stringent EU competition law dated back to earlier times, however. The European Commission's reaction to the crisis in the early 1980s had already been to significantly restrict the ability of governments to provide individual companies with financial support.[113] This policy was paired with an attempt to create incentives for innovation, something that, in accordance with the logic of global financial capitalism, increasingly shifted towards financialization. Instead of mutualizing company losses, the LeaderSHIP 2015 strategy paper called for the development of innovative financing and guarantee schemes for shipbuilding. The European Investment Bank was to take a leading role in financing the construction of new ships, both before and after delivery.[114]

The stories of the two shipyards Gdynia and Uljanik can thus be understood as a kind of Polanyian countermovement, the aim of which was to attenuate the effects of the great transformation, which began in the 1980s, by means of state-financed safeguards. That is, until the EU came onto the scene as a new quasi-state actor and began implementing new rules, proving to be considerably less attentive of the shipbuilders' narratives than the national governments were. Up until this point, the safeguards provided were, even after 1989, based on the subsidy practices of late socialism and the corresponding strategies of legitimation, on a specific moral economy, for which profits were of less importance. The fragmentation of society and the economic and political chaos that characterized the new post-communist regimes, themselves still in need of legitimation, helped the shipyards, who put a lot of work into legitimizing their demands and developing a polyphony of arguments, to assert their interests in the political sphere. Given the way in which the governments supported the shipyards and in view of the latter's rhetoric of legitimation, the political upheaval of 1989 did not in fact mark a system upheaval. Indeed, in this case, transformation appears to be more a path dependency than a structural break.

Had it not been for EU accession, this game and its lack of clearly defined end goal might well have gone on forever (or at least for as long as taxpayers were ready to foot the bill). It was therefore no coincidence that both shipyards were liquidated around five years after their respective countries joined the EU. The EU's competition policy imposed on the new member states, at once deregulatory and regulatory, put an end to the pendular movements between the state and the market that had been politically and socially negotiated. The policy was at odds with the postsocialist subsidy regimes that saw industrial and social policy as two sides of the same coin, agreeing and acting in accordance with the principles of an (in the Polanyian sense) "embedded" assessment of unbounded companies. It was not until EU accession that the "great transformation" that broke up that structure occurred, a process during which companies were largely surrendered to the market and removed from their political, and thus potentially also democratic embeddedness.

Thus, the European competition authorities put more than just two companies at risk. Through the measures it took, the nation state protected not only the holistically organized and oriented company, but also the congruence between workers' life stories, modes of production, and social lifeworlds. Preventing the state from providing subsidies thus caused all this to break down. In the next chapter, we will show how this holism, which shaped industrial modernity in Eastern Europe, just as it had done during the era of Fordism in Western Europe, was washed away. Accession to the EU thus also called a specific type of company-based community formation into question. And it is here that we shall leave the stage of high politics and management decision-making along with the knowledge found primarily in documents, turning instead to the perspective of the workers themselves.

4 Welded Together: Community Building in the Shipyards

In those days, all the schools, crèches, and nurseries were funded by the Paris Commune Shipyard. If anything broke in these schools or nurseries, workers from the shipyard's repair department would come and fix it. Because the shipyard took care of these things.
      Interview with Jan Gumiński, trade union leader from Gdynia

Uljanik owned, co-owned, or was behind virtually everything that was built. That included the marina, the stadium, the sports clubs, the cultural associations – there were a million different things connected to Uljanik.
      Sale Veruda, KUD Idijoti punk rock band

Gdynia's big dry dock was officially opened in the mid-1970s. This was a decade that began in December 1970 with the violent repression of protests led by the dockworkers in Gdańsk, Gdynia, and Szczecin and culminated in the massive wave of shipyard strikes in the summer of 1980 that spread across the entire country. In other words, the shipyard workers were not exactly thankful to the regime for their preferential treatment, instead assuming the role of the avant-garde of the working class during the protests and strikes of 1970 and 1980. This mobilization was hugely successful, leading to the creation of the independent trade union Solidarność, which quickly grew into a mass movement with ten million members (in a nation of 38 million), among them virtually every single shipyard worker. In December 1981, the regime once again resorted to violence, declaring martial law and ordering the troops to open fire on the demonstrators, including in Gdynia. These historical events, in which the workers of the Paris Commune Shipyard played a prominent role and lived up to the company's name in unexpected ways, shaped the socialist workers' experiences and frustrations.

At the same time, during the interviews we conducted with them, the former workers also expressed a certain nostalgia for the caring and nurturing shipyard they experienced under state socialism. In this respect, there is clearly something of a disconnect between the participation in these "anti-communist" protests and the rejection of the "system," on the one hand, and the fond emotional memories of the welfare services the company offered and the strong community relations, on the other. This discrepancy is even more striking when looked at through the prism of an individual worker's life story. For instance, let us take the case of the Polish shipbuilder quoted above who actively participated in the protests in 1970 and 1980, even suffering a gunshot wound, and became head of the "post-communist" trade union in Stocznia Gdynia in 1992. Yet today, this very same worker still emphasizes the positive sides of state socialism.[1]

In this chapter, we endeavour to understand these apparent contradictions and resolve the discrepancies. We will argue that the different views are neither an indication of cognitive dissonance nor one of those tricks that our memory is wont to play on us. Taking the shipyard workers' memories as a starting point, we will demonstrate the importance of solidarity and illustrate the emotional ties to the company. Crucially, it was these ties that enabled people to identify so closely with their work at the docks and with the shipyard as a community, while at the same time creating the social bonds that empowered those same workers to express protest and resistance. We will therefore examine the (infra)structural basis of the – undoubtedly ambivalent – feelings and sense of identity that were brought about and fostered by late- and post-socialist companies. The main sources we use for this are verbal accounts given by shipbuilders from Pula and Gdynia who shared their stories, experiences, and emotions with us. We collected these accounts using ethnographic research and oral history. Owing to the structure of the workforce, the accounts came mainly from men, although women were also interviewed with a view to reconstructing the clear differences that existed in the opportunities available to all genders at that time. In order to analyse these narratives, we triangulated them with written texts, such as articles from company newspapers and archival documents.

When interpreting the narrative, it was important for us to bear in mind the time when our research was conducted. In Gdynia, for instance, we were looking at a shipyard that had not existed for some years and our interviews were mainly with pensioners. When it came to Uljanik, in contrast, the shipyard was still operational at the start of our project but soon faced its final crisis. This caused a growing distrust among the managers towards prying scientists and often extremely pessimistic assessments on the part of the workers. As is typical of oral history, the research team was thus confronted with two temporal levels: a distant level that the memories we captured referred to as a subjectively experienced history, and a second level that focused on the present, that is, the time of study. In both contexts, our interlocutors, and thus the authors, too, found themselves in "stormy waters" that brought some of the different sediments of the long and profound transformation to the surface and washed others away. Besides the ubiquitous topos of decline, we gained significant insight into the lost forms of communality and modes of imparting meaning. The way our interlocutors – but also we ourselves – interpret the past is marked by experiences and discourses that come later. It is not, however, only the aggregate experiences and the respective present that shape our interlocutors' opinions – the reverse is true as well. How someone perceives the present is ultimately also related to their life story. In the end, when we analysed our field research and interviews, we were in fact operating on a third temporal level – that of the scribe's quarters, our office, which is assumed to stand on solid ground. Yet, even here, the never-ending transformation of the research subjects continued to have an effect.

So far, this book has documented the structural conditions, companies' attempts to adapt to ever-changing conditions, political intervention, the challenges of the world market, and the economic costs associated with the protracted survival of the industry giants in Gdynia and Pula. We argued that the state socialist principle of holistic manufacturing enterprises failed because of its inability to achieve the flexibility required by modern capitalism. However, this was not the only friction that transformation created for a holistic understanding of labour: there was also a gradual breakdown of the holistic lifeworlds that had evolved during socialist modernity when the shipyards were more than just a workplace. In fact, the employer organized much of the workers' lives, whether directly or indirectly. They provided social services and other resources that were essential for the social infrastructure under state socialism, but also for some years beyond. Notions of what constitutes a good life and a fair society are still shaped by those years to this day.

Socialist production aimed at integration was mirrored in the workplace-centric organization of society. Work and community seemed to be inextricably linked. This was not dissimilar to the set-up in the West's typical Fordist industries, with one crucial difference, however – by the 1970s at the latest this connection had disappeared in the West as a result of the structural adjustment to the new economic era following the oil crisis, while in socialist countries it remained instrumental.

"There was thus a material and affective dimension to the gathering of bodies," observes Andrea Muehlebach in (post-)industrial Sesto San Giovanni, a suburb of Milan.² The socialist factory floors had an additional dimension to the experiential world of Italian industrial labour. In an economy of shortage, with machinery that often malfunctioned, it was impossible to fully implement Taylorist and Fordist methods to control the workers' bodies and movements.³ Under socialism, the workers often had more control over their work, as production depended very much on their resourcefulness and ingenuity when dealing with unforeseen problems.⁴ That said, the gender dimension was apparent here, too. Workers in the male-dominated heavy industry sector enjoyed far more leeway than those in light industry, including assembly line production, where a particularly large number of women were employed. Given the opportunity, female workers in Istria thus did not hesitate to move from the sardine canning factory in Banjole to the nearby shipyard in Pula, which offered better working conditions and higher wages, as one female worker describes in an article in the Uljanik company newspaper.⁵ The improvisation that was frequently required at work strengthened the cooperation on the factory floors, reinforcing the bonds between the employees and their sense of community. This type of autonomy in the workplace was especially pronounced in the shipbuilding industry, which is marked by production models typical of skilled trades, long production cycles, and sprawling facilities.

The labour-centric socialist lifeworlds led to expectations of the postsocialist regime that, as our interviews illustrate, reality often made a mockery of. The new management cut down on the community functions, concentrating instead on what was known as "core business." In line with the principles of modern business management, these reforms aimed to shift control of production from the grassroots to the managers, a move that also undermined the very foundations of factory floor solidarity. These processes were gradual, however, meaning, long after 1989, the shipyards were still more than just a place where ships were built. It is precisely this partial resilience inherent in some community functions and the local endeavours to preserve these very functions that are far more evident in our perspective "from below" than in official records, where the government and company management perspective is more prevalent.

But what was or in fact is at the "core" of these companies? For many of our interlocutors, it was not only, or not so much, a question of generating profits but in fact far more about creating a community and welfare as well as identity and meaning. Social anthropologist Elizabeth Dunn made similar observations during a visit to a privatized food processing plant in Poland in the 1990s. The female workers she studied emphasized that "production is a relationship between people" and work is "an action by one person for another."[6] Socialist production was inefficient when it came to manufacturing things, but incredibly successful at producing social relations. In *Geschichte und Eigensinn* (*History and Obstinacy*), Oskar Negt and Alexander Kluge describe their complex search for the logic behind capitalism:

> It can be said that capitalism is the mass production of goods in which people are incidental. Socialism is the mass production of relations between people and with nature in which the production of goods is incidental.[7]

On the company level, this implied a different form of bookkeeping where the expenses of the socialist enterprise were offset against the collective welfare gains. Incidentally, Karl Polanyi advocated this, on an abstract level, in 1923, when, in an essay on socialist accounting, he called for social services for the community to be incorporated into companies' balance sheets and for businesses to be measured not only by their production output.[8] The name of Stocznia Gdynia under socialism, the Paris Commune Shipyard, was thus very apt.

On the following pages we investigate the transformation of the community-building purpose behind production from the perspective of those who actually experienced these changes, resisted them, and adapted to them. We chart the accompanying perceptions, feelings, and meanings, and we identify interwoven social relations that help us re-evaluate the "absence of productivity under socialism" and its persistence beyond the political transition. The communist regime quite patently failed to achieve its utopia – the frequent protests by the workers and their subversion in the workplace, whether overt or covert, were important factors here. But even during these protests, relations of solidarity developed on the shop floors, reinforced beyond the factory gates through the wide range of recreational and cultural facilities offered by the shipyard. Consequently, as argued by sociologist Michael Burawoy, through the strong company-based community and the relations of solidarity it cultivated, the state socialist production regime created a kind of counter-socialism "on a small scale" – in other words, in the workplace.[9] Underpinning these relations of solidarity was, among other things, the cultural infrastructure that the firms created with a view to perfecting the "socialist personality" and enabling people to become a new kind of human being – the results of which often turned out quite different than the cultural policymakers had intended.

The Subculture of Uljanik and Stocznia Gdynia as a Food and Leisure Cooperative

Socialist firms invested considerable resources into "cultural work." This served the "civilizing mission" of socialism, which aimed to create a "new man."[10] These cultural activities ranged from dreary propaganda events to support for creative hobbies – drawing, writing, and theatre classes were organized, or a venue provided for people to play chess or musical instruments, for instance. The pages of the Uljanik company newspaper were full of information about the wide variety of events organized by the shipyard's own cultural club, which was known by its acronym KUD (Kulturno umjetničko društvo). The KUD, which had been founded in 1967, offered activities such as readings, folk dancing, and folk music. The literature section of the KUD published poems and other literary works written by its members – partly to bear witness to the creative potential of the "ordinary" workers.[11]

While the club's program in many ways conformed to the party's ideas of the "self-optimization" of socialist citizens, it also offered space for anti-hegemonial discourses. The sociability linked to the workplace as well as the capacity for cooperation comprised not only conventional welfare and cultural institutions but also subcultural spheres. The latter, in particular, successfully contributed to community and identity building because they were situated outside the official ideology and allowed more self-determination. This is evidenced by Uljanik's best-known home-grown cultural talent – the punk band KUD Idijoti, founded in 1981. Even the intentional misspelling of the second part of the band's name (in standard Croatian, the word would be written *idioti* – a detail that well-meaning editors frequently "correct") conveys the counter-cultural message. Sale Veruda (real name Saša Milanović), who provided backing vocals and guitar and to this day still performs with his punk band, Saša 21, describes the significance of Uljanik for the band:

> KUD Idijoti was a quartet. Three of its members worked for Uljanik: the drummer, Tusta, and me. The base player was a concierge at the municipal hospital around the corner. This meant our working hours were basically stable – we mostly worked the early shift from 7:00 a.m. to 3:00 p.m. This meant we had afternoons to practise and work on our music, and we really practised a lot. We talked about our music a lot, about other bands, about the situation on the world stage: we followed everything. In other words, there was no stress. Nowadays, it is much more difficult because four or five of the band members might be working different shifts. One

is on early, another is working an afternoon shift, the third works in shipping and is away for a few days. Then comes the tourist season and people work the whole season – June, July, August, September. So some bands can't even perform anymore these days …

We used to meet there, in our workshops – he [Tusta] worked for TESU, the factory for electrical machinery and appliances. Our carpenter's workshop was close by, which meant we could sometimes get together – especially later, once we were in the band together and we needed something, we could meet, you know, during our half-hour lunchbreak.

The Uljanik shipyard was thus closely interwoven with Pula's subcultures; its work rhythms and routines created space for creativity, and the jobs it offered provided the artists with social stability. The Yugoslav public saw the band's links with Uljanik in a positive light. In Sale's words, the shipyard "lent the band an element of authenticity, so when it came to socially sensitive topics and lyrics, people knew that we weren't just shooting our mouths off but that we had really lived and experienced this stuff ourselves." Sale pointed out, however, that they were still not allowed to perform in the shipyard's club: "Since the 1970s and 1980s, they had a policy of only inviting the Yugoslav A-list if you like. The top bands. A small band like KUD Idijoti had no chance of ever getting to play there, even though we were a local group."

Even before KUD Idijoti's time, Pula was renowned throughout Yugoslavia for its music subculture. The company-run club was one of the top venues for Yugoslav rock music for over fifty years and enjoyed cult status beyond Istria. The club was located near the shipyard's main entrance, up on a hill close to the city centre (figure 4.1). Since the 1990s, it has operated as a night club but no longer under the auspices of Uljanik Standard (the subcontractor for social services). Until its closure in 2022, it was run by private leaseholders, albeit under the same name (in 2023, the building was transformed into a burger restaurant, another telling example of the nature of post-socialist transformation). In other words, the shipyard and the music venue, originally meant for the shipyard workers, put Pula on Yugoslavia's pop and subculture map. To this day, the link between Uljanik and the music subculture continues to cultivate a sense of community (*zajednica*) that is manifest through a network of cooperative relations across the entire city, and the wealth of positive memories that this conjures up.

In Gdynia, too, the shipyard created a strong community by acting both as a provider and an entertainer in some quite surprising ways.

Figure 4.1. Rock Club Uljanik, 2021
Source: Photograph by Ulf Brunnbauer.

A member of our research team (Piotr Filipkowski) met Jadwiga and Leszek in their small but beautifully renovated cosy apartment in one of the former dockworkers' apartment blocks located in the Obluże district of Gdynia. The couple gave Piotr a warm welcome, serving him tea and cake, clearly delighted about his interest in "their" firm. They had worked for the company their entire lives, something they were enormously proud of, and showed Piotr family photos, many also featuring the shipyard. One of the pictures was of the couple's entire family, this time not in front of the socialist shipyard with the unusual name "Paris Commune," but rather in front of the successor company, Crist, where their son had found a job. This shows the real dynasty of workers this family was, something that, much like in Pula, is seldom the case in Gdynia today.

Leszek begins by telling us about his father, who had come to Gdynia from a tiny village in the region to work as a welder. After a number of years, he was promoted to the position of senior foundryman, heading a team in the hull department – in other words, he worked at the very heart of the production process. The way Leszek tells the story, it sounds as though all the ships ever built in Gdynia passed through his father's hands – as though they were his personal creations. In Leszek's account, ships seem to be more of a product of craftsmanship (and shipbuilding even an art form in itself) than the result of large-scale heavy industry processes. And yet, despite his clear respect for manual labour, Leszek in fact chose a different career path to his father. He attended the vocational college set up by the shipyard and was offered a position in the toolroom. After a few years, as an activist in the Socialist Youth Association (Związek Młodzieży Socjalistycznej, ZMS) – the leading mass youth organization in communist Poland – he was put in charge of the shipyard's recreational and cultural activities. The couple reminisce about Leszek's time in this position, which is also when they met:

> Youth organization – what we did here was organize activities for young people outside working hours, both in the company itself and elsewhere.... After work, you know – fun, recreation, trips, excursions. But other things too ... theatre and cinema trips ... We had our own youth club at the shipyard, on Śląska Street. It was very impressive, the Frigate, it was called. It was an amazing thing.... We organized sports activities on the shipyard's grounds.... We had lots of friends and work colleagues, it was wonderful. I have such good memories of those days.

In these memories of the 1970s there is a congruence between biographical and structural factors. Our interlocutors were between twenty and thirty years of age at the time and were just embarking on their professional careers. Leszek quickly advanced in his career, progressing from a semi-skilled worker to a full-time activist within a short space of time. While his father, a tradesman and skilled worker, built the biggest ships Poland had ever seen in the shipyard's recently opened dry dock, Leszek was climbing the next steps of a very different career ladder, parallel yet (seemingly) disconnected from shipbuilding. Leszek continued with his training – first earning a qualification from the shipyard's technical college and later graduating with a business degree from the University of Social Sciences in Warsaw, an elite school for members of the party and the government.

During the interview, Leszek depicts his support for the socialist state neither as faith in some abstract "system" of sorts nor as a sheer pragmatic "necessity," as is often the case in oral history interviews in post-socialist countries. Instead, he interprets his political activism as a service to the shipyard community. His career path, which, although in many ways very different from his father's, was similar in terms of how they identified with the shipyard, shows the expansion of the firm's non-production-related functions. This meant that the shipyard was not just one of the actors involved in the industrial modernization of the country but also a key institution in the socialist welfare system. The arrangement promoted a certain lifestyle among the workers, who were encouraged by people like Leszek to spend their free time in a more productive manner.

When the Polish economy toppled from the peak of Gierek era modernization to the doldrums of the economy of shortage of the 1980s, Leszek's area of responsibility changed quite dramatically. While still responsible for the non-production-related functions, his main focus was meeting the basic needs of the workforce, providing consumer goods such as food and fuel:

> In these difficult times, the shipyard traded in almost anything.... Yes, there were people who built [ships], but we bought cigarettes, socks, coffee ... onions, and what was really important – potatoes. Potatoes and onions were my main occupation! In those days, as an employer the shipyard was required to provide its workers with potatoes and onions. There were also apples ... Vouchers were distributed, all that kind of thing, the whole bureaucratic apparatus. And I was responsible for supplying the entire shipyard, the whole workforce ... Tonnes, hundreds of tonnes, were brought in. We bought these potatoes from different government cooperatives [Państwowe gospodarstwo rolne] and independent farmers, it varied ... For onions we travelled to Inowrocław,[12] we brought back entire cars full, 20 to 30 tonnes. This was all delivered to the shipyard and then we used our cars to transport it all from the shipyard to the workers' homes ... We had to get the potatoes to the workers!

This example illustrates not only the consequences of the "economics of shortage,"[13] but also a fundamental feature of social welfare under state socialism: many of its services were provided through companies. As a result, it made a huge difference to an individual's standard of living whether they worked for a firm that was capable of organizing things properly – or not, as the case may be. Leszek, who just a few years earlier had encouraged people to use the benefits of the "system," was now just as committed to helping them navigate its fundamental flaws.

One of the reasons for choosing the passages quoted above was that this single example perfectly illustrates the entire gamut of social emotions that could evolve from a work-related lifeworld. Today, Leszek summarizes these emotions in one simple sentence: "I have such good memories of those days." What makes these memories "good" is not (only) people's nostalgia for their youth and the pride they felt at being a shipbuilder from Gdynia but also how proud they were to have represented the shipbuilding community in all kinds of areas. Sociologists have often described Polish society under socialism as a "social vacuum" where the meso-level of social integration between very close family relations on the one hand and the state on the other is missing.[14] The example of the Paris Commune Shipyard clearly shows that this vacuum could be filled to a certain extent – with social bonds but also with the company-level privileges, as reflected in Leszek's life story.

Despite all the repercussions it had for the shipyard, the political transition that began in 1989 did not bring these social practices to an abrupt end. Rather, the significance of these practices began to wane gradually. In the 1990s, there was no longer any reason to provide consumer goods as the economy of shortage was replaced by individualized consumption. Although there were still cultural and sporting activities, these were now organized on an individual basis. As a result, Leszek's job changed, too, assuming a more prosaic character. In the 1990s, for instance, his function was to supervise cleaning operations, in particular the disposal of production waste as well as the workers' everyday rubbish.

The Multifunctionality of the Shipyard: Housing, Sports, and Welfare

The Uljanik shipyard is situated in the Bay of Pula, immediately adjacent to the historical old town, stretching away from "Olive Island" along the southeast side of the bay to a small repair shipyard and cement factory in the west (see map 4.1). If you walk over the hill, southwards away from the shipyard, you will reach the districts of Stoja and Veruda. It is these parts of town that are home to the biggest population of Uljanik workers (map 4.1). This is also where you will find the athletics and football stadiums, once built by the shipyard and still proudly bearing its name. For the people who live there, this is a symbol of the good old days.

The stadium was built in the mid-1980s, the former director of Uljanik Standard tells us, "because Uljanik had so much money that [the factory management] did not know what to do with it, so they just decided to use it to improve the quality of life of those living in the city. To satisfy people's recreational needs" (U.S.). Until 2011, the stadium was the home ground for the biggest local football club, NK Istra 1961, which was the product of a merger between the Uljanik factory club and the city-funded NK Pula in the early 1960s. These days, on warm spring evenings, you will see people from the neighbourhood out for a jog or working out on the track. The stadium is also used for training by foreign students attending a sports week and the football field is where

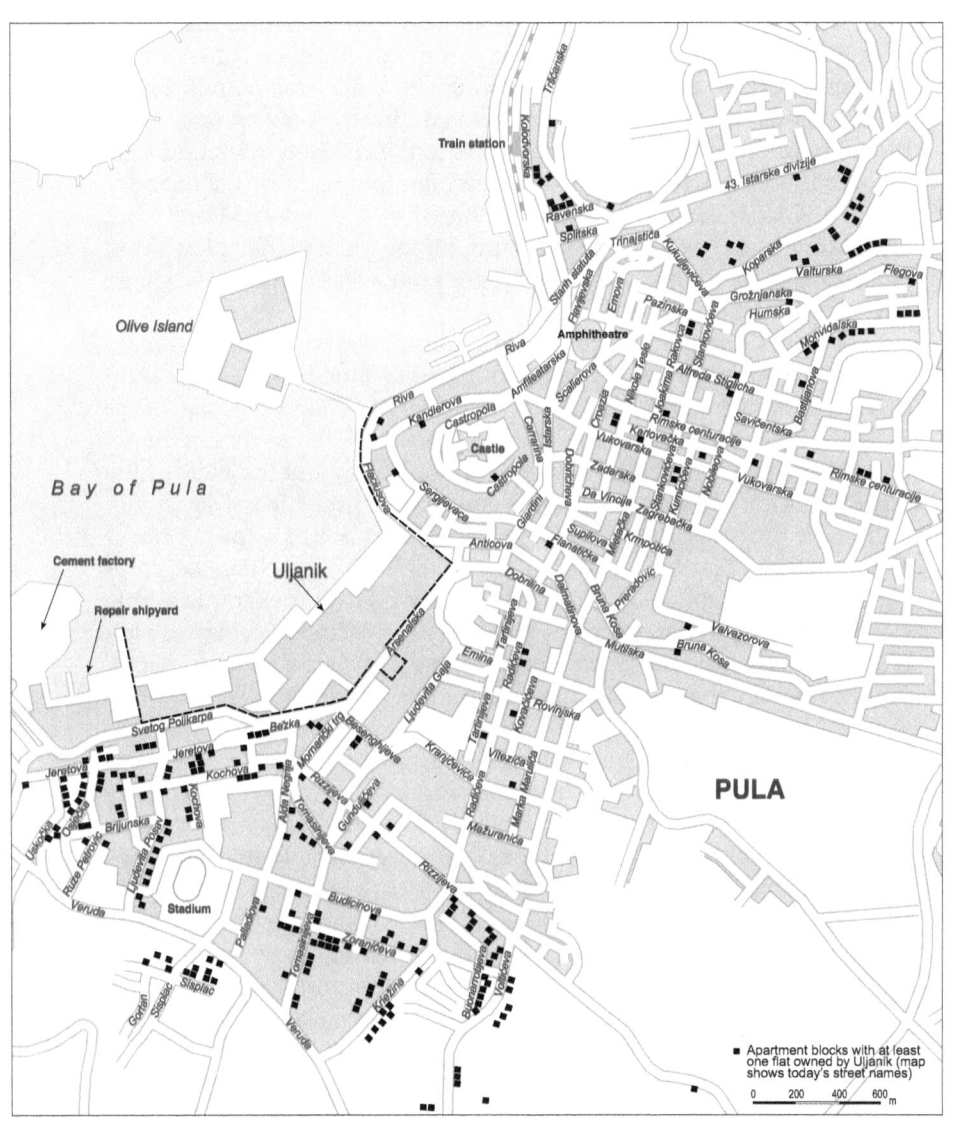

Map 4.1. City plan of Pula indicating apartment blocks and houses with company flats, 1991

Uljanik's senior and junior teams both hold their regular training sessions. Behind the Uljanik banner, there are also a couple of basketball courts, where children often play. As sport is so popular in Pula, it was important that, despite its decline, the shipyard continued to support amateur sports to the very end. The club run by Uljanik Standard, NK Uljanik, entered the district league and continued to be sponsored by the shipyard, playing all its home games in the stadium in Veruda. One shipyard worker who is an avid football fan told us that "the majority of the NK Uljanik players work for the Uljanik shipyard. They work from Monday to Friday but get nothing for the football they play." That is the deal: "We give you a job, you get your wages, and in exchange you play for free."[15]

Less than a kilometre south of the stadium, the popular Lungomare coastal promenade winds through the forest to the beach. A few hundred metres to the east are the tower blocks of Vidikovac – also originally built for the families of shipyard workers. The apartment blocks are certainly not particularly aesthetically pleasing but the view of the Bay of Pula and the Brioni islands is stunning. A little further south you will reach the forested Verudela peninsula that boasts a tourist resort and Bunarina Marina, including the Uljanik diving club, the shipyard's rowing clubhouse, and a great many boats, for example a little steamboat that takes passengers to the offshore island of Fratarski otok (Franciscan Island, also called Veruda Island). In the past, many Uljanik families would spend the summer camping on the little island while they let their apartments out to tourists.

All these places once belonged to the shipyard but when the company went bankrupt in 2018, the ownership structure changed. The stadium in Stoja was sold to the city, which now maintains it for general use and has even added a big indoor pool. Fratarski otok, an island once partially owned by Uljanik, now also belongs to the local authority. This transfer of ownership echoes the social and economic changes at the time, in particular, the shipyard's waning influence. These shifts enable us to map just how the meanings and value regimes associated with Uljanik have changed over the years, as have the social relationships and power structures underlying the system of ownership.[16]

It is already a well-known fact that under state socialism companies played a pivotal role in social welfare and cultural life. Full rights to civic participation were acquired through the workplace – and the quality and quantity of social services you enjoyed depended on the economic and political clout of the company you worked for. By tying welfare to the workplace, managers and economic planners hoped to commit the workers to their employer. This can be explained by the fact that socialist economies were characterized by significant movement of labour, all the more so in Yugoslavia because of the opportunities to move to the West.[17]

In Yugoslavia, the system of self-management strengthened this corporatization of welfare in two respects. First, company taxes were the main source of financing for municipal services, including retirement homes, childcare establishments, cultural activities, road maintenance, fire services, and so on – all of which were organized on the basis of "self-management." In 1979, for instance, some 50 per cent of all the local taxes collected in the city of Pula that year came from Uljanik.[18] Deductions from the workers' gross wages not only comprised taxes and contributions (*doprinosi*) for "regular" social security (accident and sickness insurance, as well as pensions), but also charges for the municipal culture and social budget, the child protection agency, municipal social services, as well as primary and special needs schools. Depending on the place of residence, this could make up as much as 29 to 30 per cent of a person's gross salary.[19] Through the delegate system, Uljanik deployed representatives to a wide range of "self-managed interest communities" that were responsible for the organization of welfare and cultural work in the city and decided how the funds should be distributed.[20] In other words, the system of self-management created countless formal – as well as informal – interconnections between the shipyard and the city. The result was that the company's external boundary became fundamentally blurred and the city's influence, in turn, extended deep into the shipyard, leading to mutual dependence.

At the same time, the shipyard directly provided its workers and their families with welfare and leisure services. It maintained a certain "standard of living" (*životni standard*), a term that was frequently used by the management and trade union alike. Since the division of Uljanik into the basic organizations of associated labour in the mid-1970s, a "working group for common services" has existed for this purpose. This was not one of the basic organizations, but rather a service provider for occupational health and safety, military self-defence, personnel tasks, and legal issues. The working group had a long list of expenses, including funds for campaigns and activities on special holidays (the "day of children's joy" and "Uljanik day," for instance), for awards and prizes, schoolbooks and training, obituaries to late members of staff and retired workers, retirement parties, nurseries and crèches, sports clubs, and so on.[21] It was therefore not only about the individual workers' standard of living but also about the "social standard" at large (*društveni standard*), something the shipyard felt equally responsible for, as the company magazine never tired of pointing out (see figure 4.2). During the course

Figure 4.2. Components of the "social standard" depicted in *Uljanik*, 1986
Source: Courtesy of the Historical and Maritime Museum of Istria.

of the shipyard's restructuring, once Yugoslavia's self-management socialism was no more, in June 1990, this working group became the subcontractor Uljanik Standard. The latter remained responsible for both social services and leisure infrastructures as well as sponsorship for the company's sports clubs. During the major economic crisis in 1990–1, it also helped procure food and fuel at reduced prices.

Uljanik Standard also managed the company apartments. During an interview we held with the former Uljanik Standard director, he repeatedly referred to the housing question (*stambeno pitanje*) and housing needs (*stambene potrebe*). He had worked for the shipyard since 1984 and stressed that housing needs had always been central for the working group. Data from the company archive show that from 1981 to 1985 alone, 540 new apartments were built for Uljanik employees and that in 1985 itself, another 100 were under construction.[22] In 1992, the shipyard owned 2,744 apartments – in a city that at the time had around 50,000 inhabitants (for the spatial distribution of the apartments, see map 4.1).[23] Moreover, the shipyard provided low-interest loans for the construction of detached single-family homes. A worker we interviewed recalled one of the more recent housing construction initiatives:

> It was in the early 1980s, around '83, '84, or '85, I don't remember exactly when, Uljanik built two brand-new roads in Vidikovac, prepared all the land, set them up with water, electricity, telephone lines, paved the driveways, and anyone who had social points [the number of points depended on the number of children, length of service, and need] was able to get a plot of land instead of an apartment. They got excellent loans to build their houses – there were prefab houses but also solid masonry ones, but these belonged to a cooperative [*zadruga*]. Everyone got the materials they needed to build their houses for a decent price. So we showed solidarity, we were all friends, and we all helped. (Č.Č.)

The workers vaguely remember the waiting lists and the payments necessary to get one of these sought-after apartments. In the 1980s, "payments for housing construction were deducted from our wages each month, 0.5 per cent or 1 per cent, I don't remember anymore." At least "everyone who came to work for Uljanik could count on getting some kind of accommodation from the company within five to eight years" (Č.Č.). One fierce local dispute in March 1989 shows just how aggressively and assertively Uljanik ran the whole residential construction part of the business. The municipal authorities accused the shipyard of building without a permit, which the Uljanik representatives did not even attempt to deny. Instead they explained that this was unavoidable because the city was incapable of issuing building permits on time. Bruno Bulić, chair of the trade union committee, presented the shipyard's position: "We spend a lot of money on this city and we will continue to do so.... All we want is to be part of life in this city in the future, nothing more, nothing less."[24] Thus, social investment helped not only with community-building, but also with the accumulation of social and symbolic capital, which could be used to make demands of the political authorities.

In the post-socialist era, when the tasks and obligations of Uljanik Standard were gradually reduced and the workforce scaled back, the apartments were privatized. The shipyard also urgently needed the cash from the sale of the apartments to avoid bankruptcy. According to the former Uljanik Standard director, the sale of the apartments began in 1992 and was largely finalized by 1996. Only a small part of the company's inventory of apartments remained unsold because the tenants were either too old or could not come up with the necessary funds to buy their apartment. For the vast majority of tenants, the privatization of the apartments presented a unique opportunity to acquire ownership of residential property at an affordable price – they had the option of repayment over a long period with small monthly instalments (some were even still paying small monthly instalments at the time of writing). The sale of the apartments proved to be an important contribution to social stability at a time when wages were very low – and also an investment in the future, as the owners now let their apartments to tourists in the summer months.

Irrespective of who has the property rights now, the once company-owned apartment blocks, the stadium, and the remaining sports facilities continue to embody the strong presence of the shipyard in the urban landscape. Not only are they powerful *lieux de mémoires* of industrial modernity in general and of life with Uljanik in particular; they are also widely used and have thus become socially appropriated living spaces. In the cafés between the apartment blocks where the former Uljanik dockworkers regularly meet, more often for a beer or *gemišt* (wine spritzer) than for coffee (other genders tend not be part of these groups), the community of shared memory is kept alive day after day. Thus, during our field research, the memory sites served as a connection between the past, the days when Uljanik was still operating, and the present – albeit a present that was to be frequently taken apart and rebuilt.

When it comes to multifunctionality, the publicly owned shipyard in Gdynia was in no way inferior to the self-managed company in Pula. Although these expenses were high, they were considered part and parcel of the company's running costs and were not up for discussion. As a rule, besides production costs, the Paris Commune Shipyard's annual reports also included "indirect costs" and the wide range of welfare services provided were entered in the books under this position. For example, this is how they recorded the holiday organized for 4,718 members of staff and their children in the company-owned sanatoriums and holiday resorts in Wieżyca, Szklarska Poręba, and Jastrzębia Góra in 1980. The shipyard also maintained cultural and educational facilities, such as a library and highly subsidized early childhood education. A lower-cost item on the books was the allotments that were extremely popular in the socialist economy of shortage. In 1980, total spending on "social activities" amounted to 31 million złoty. Only a fraction of this sum came from the proceeds of the activities themselves; the lion's share (20 million złoty) was covered by the shipyard's social fund.[25]

Besides the social activities, the annual report also included "household activities." Here, the costs for actual residential construction in 1980 were as little as 3.2 million złoty compared to operating costs as well as the accommodation for additional employees in workers' hotels and other lodgings, which were by far the biggest budget item at 55 million złoty. This same category also included the costs of the workers' medical treatment in the company health centres as well as the cost of feeding workers in eight dining halls, six canteens, seven food kiosks, one fruit and vegetable kiosk, one grocery store, one soup kitchen, and a tea and coffee kitchen. The company's vocational colleges were also classified as "household activities," as were the services provided by the purchasing department. In 1978, for instance, 925 tonnes of potatoes, 60 tonnes of onions, and 100 tonnes of fruit were distributed to the company's employees and pensioners (as described by Leszek above). In 1980, "household activities" cost the company a total of 85 million złoty, meaning they accounted for around 15 per cent of the shipyard's net loss.

In the 1990s, Stocznia Gdynia's "social" and "household activities" were also restructured, although, much like the ownership transformation, this process, too, remained incomplete. The welfare service providers were converted into a separate company, but in practice nothing much changed. One example of this is the sanatorium and holiday resort in Wieżyca, a village in the Pomeranian lake district, where – despite outsourcing – shipbuilders remained the most important group of customers. The same applied to the Bałtyk sports club, which, in spite of being split off, still recruited its members from among the shipyard workers. Testimony to the sheer resilience of this multifunctionality is the wide range of services offered by the hotel and tourism company founded by the shipyard:

1. Provision of tourism and recreational services, 2. Health resort services, 3. Provision of food and beverages as well as other catering services,

4. Lease of business premises and residential buildings, 5. Warehousing services and property management, 6. Food production and processing.[26]

The unbundling of the different parts of Stocznia Gdynia was thus just as leisurely a process and to some extent also as purely formalistic as it was in Pula.

In Gdynia, one area where the restructuring efforts did in fact set a substantial transformation in motion on the local level was the health care system. In the year 2000, the company-owned hospital, which was located in the shipyard itself, was sold to Euromedicus, a subsidiary of the national health care fund, which coordinated and financed health care provision in Poland. Uljanik did not have its own hospital, as the Yugoslav system of self-management was designed more for companies to co-finance self-managed providers of welfare and health care services. Consequently, shipyard employees lost their privileged status in hospitals, with many of them soon complaining of waiting times of several months. The situation for employees of the legally independent subcontractors was even worse. Euromal, for instance, a company that offers ship painting services, signed a contract with a private health care service provider that meant that their employees were no longer entitled to be treated at the former shipyard hospital. This resulted in a very serious problem when one of their workers got metal shavings in their eye in a work-related accident:

> As was standard practice in such cases, I went to the eye specialist at the shipyard. But I was in for an unpleasant surprise. They refused to treat me, despite the fact that I am insured! At reception, I was told that Euromal employees are no longer allowed to go to [the Euromedicus] hospital in Czechosłowacka Street. I was sent to the hospital on Chrzanowski Street. But I had to wait until the next day and pay 30 złoty. It felt all wrong to me – I am a voluntary blood donor and have worked for the shipyard for over 20 years.[27]

Quite coincidentally, this memory also demonstrates the value of oral history and its connection with written sources – it was only thanks to this verbal account that the research team became aware of this issue at all, prompting us to look for more evidence in the local press. And indeed we located a critical article from June 2004 to which the Stocznia Gdynia ombudsman replied that the shipyard "now had much more serious problems ... than the hospital."[28] The management conceded that, given the shipyard's ongoing liquidity problems, it had prioritized paying outstanding wages and saving jobs over health care services.

This was counterproductive when it came to the shipyard's image and its position in the social fabric of the city, despite the obvious positive impact that the health care savings had had on Gdynia's balance sheets.

Like in Pula, housing was a key component of the welfare services provided by the company in Gdynia, too; and fundamental changes were also seen here from the 1990s on. In 1970s Poland, the state and the local authorities had gradually moved away from centralized housing construction. Based on a new law on cooperatives, the Paris Commune yard founded the Stoczniowiec (Shipbuilders) housing cooperative, which was headquartered in one of the shipyard buildings. By 1989, Stoczniowiec had built over forty apartment blocks using preferential loans from PKO Bank. And, up until 1995, the members of the cooperative were able to buy one of these apartments at favourable conditions – an opportunity many shipbuilding families seized. After this first period of transformation, the cooperative was legally separated from the shipyard, moved from the shipyard premises, and has since been selling properties on commercial terms to wealthy customers – a group that, as one of our interlocutors bemoaned (B.W.), rarely includes shipbuilding families.

Besides housing, the provision of training and qualification opportunities were an important way of securing the workers' commitment and integrating new employees into the shipyard community. Both shipyards offered an extensive training program, they cooperated with vocational colleges or ran their own, and they financed scholarships for specialized shipbuilding programs at technical universities or – in Croatia's case – at the Faculty of Shipbuilding at the University of Zagreb. Thus, an entire ecosystem of shipbuilding institutions, expert knowledge, and personal networks evolved. Vocational training was closely intertwined with the shipyard's work processes:

> I remember being taught by Mr. Zwierzyński, the deputy director of the shipyard. Of course, a lot of engineers, a lot of people from the shipyard taught technical subjects. From our first year on, we were given practical experience on a ship in the shipyard. Most of them, like the welders, etc., were on a ship from the very start; from the second or third year they were allocated to teams. They already got to know their people. The trainees knew the workers and the foremen. And once they had finished their training, every foreman, every supervisor knew who they were and who they wanted to have on their team, etc. There was real continuity. (G.K.)

Given how successful this model was, it is hardly surprising that the education and training infrastructure was pretty much left untouched during the first period of transformation. Although Stocznia Gdynia's training department was disbanded in 1990, the shipyard continued to operate the shipbuilding vocational and technical colleges. In keeping with the socialist tradition, students attending these institutions became acquainted with the day-to-day work from the very start, even after 1989, as another worker we spoke to told us: "I started my shipbuilding training in 1989. First at the vocational college then the technical college. And from the word go, I had practical placements at the shipyard. I was connected to Stocznia Gdynia the whole time" (G.K.). Although this particular worker originally came from southern Poland and was not born into a shipbuilding family, his entire working life was spent in the industry – after Stocznia went bankrupt, he began working for a specialist company that was set up on the former shipyard premises. However, in 2004, the shipyard sold its vocational and technical colleges to the city in order to pay off some of its debts. The only shipbuilding training centre on Energetyków Street, around 500 metres away from the shipyard's main entrance, operated until 2016, when it was closed down as part of a nationwide education reform. At the closing ceremony, Małgorzata Zwiercan, a member of the Polish parliament, referred to the closure as a "symbolic act signifying the end of the process of winding Stocznia Gdynia down."[29] The situation in Pula was similar. Not least because of the dramatic decline in demand, the shipyard's ties with specialized training institutions increasingly weakened, although at this stage they were not severed entirely: bankruptcy would see to that in due course.

Under capitalism, many of the services that had previously fallen under the shipyard's "social" or "household activities" were provided by independent companies. Given their profitability, the decision taken in the 1990s to restructure and concentrate on the chosen "core business" was called into question. Could it be that the reformers had completely misinterpreted what the shipyard's core business was? Was it in actual fact "social" and "household activities" rather than shipbuilding? If this were the case, Stocznia Gdynia and Uljanik would secretly still exist today, albeit dispersed and under another name.

Coping with Collapse: Shipbuilding, Hierarchies, and a Longing for Community

When the workers recall the trade unions in late socialism, they do not necessarily mention strikes, industrial action, and wage negotiations – the common associations made in Western Europe – but rather the gifts and excursions the unions organized. Compared with the post-socialist era, the trade unions focused far more on social issues and less on labour law, a civil servant from Pula told us in an interview:

> We organized *zimnice* [pickled vegetables for the winter] for the workers, we made sure they got, say, cheap potatoes, sometimes some meat, there were excursions and preventative health retreats they could take part in. We were much less involved in labour issues [*radnička pitanja*] than we are today. (B.Bu.)

A long-serving Uljanik worker told us that the trade union "organized weekends in a hotel in Poreč," including

> medical healing treatments, thermal baths – the trade union paid for this, as long as you provided a medical certificate. It was also thanks to our trade unions – they were the best! – especially the Uljanik union, that we got a winter and summer holiday in Slovenia. (M.D.)

As already mentioned in other contexts,[30] the role of trade unions in Yugoslavia and the other socialist states (Solidarność in Poland is an exception here) was different to the one they played in capitalist countries. In Yugoslavia, the trade unions also toed the party line and functioned as a transmission belt between the party and the workers.[31] A number of our interlocutors told us that no one had a choice, "in the system back then, there was only one trade union, which you effectively had to join; if you were employed [anywhere], you more or less automatically became a member" (notes from our fieldwork). A former member of the workers' council added that he felt like he was being used as a pawn (*pijun*) in a much bigger game – and so he left the trade union.

The socialist trade union model could do little to counter the neoliberal economic reforms introduced after the end of communist rule. Given the precarious state of the economy and growing unemployment, the socialist trade union activities were strongly geared towards discipline. The pluralization of the trade union movement in Croatia after 1990, largely along the country's political dividing lines, proved to be counterproductive. The shipbuilders in Pula were ultimately represented by three different trade unions. The biggest of these – Jadranski Sindikat (Adriatic Trade Union) – was perceived as being more militant. For a long time, representatives of the other two unions (the Metalworkers' Trade Union and the Trade Union of Istria, Kvarner, and Dalmatia) sat on Uljanik's supervisory board alongside members of the shipyard's management. Some workers saw this as a conflict of interests, which, in their eyes discredited them as "yellow trade unions." They interpreted this as a continuation of the role the Yugoslav unions had played in late socialism. At that time, the constitution and the national laws saw the function of trade unions as being to implement state policies, not offer support to the workers and their autonomous organization. The worker self-management, which, under socialism, should have compensated for the trade unions' failure to represent the workers' interests, also proved incapable of preventing the growing alienation of the workers from the company management and those in power.[32]

At the end of June 2018, Ivo, a pensioner who had spent his entire working life at Uljanik, ultimately being promoted to foreman, invited a member of our research team to meet him at a café at the Veruda market. This is a very popular meeting place among Uljanik workers – indeed some of the men drinking their coffee or beer are still in their work overalls, which visibly underlines the shipyard's presence. When Ivo talks about the shipyard, he almost has tears in his eyes, and although he says several times that he does not know much about the current situation, he still repeatedly contrasts his memories with the present day. He points out that Uljanik has to have "the right people in the right positions." In his view, the foremen should be in more senior positions because they know what tasks the workers under them have to fulfil – this would enable the managers to earn the workers' respect.[33] Ivo is not the only one to suggest that the old directors had worked their way up the ladder step by step, whereas the new managers often know next to nothing about what the workers do. They are a separate class – a group of people who "came in from the outside". This shift can be interpreted as a kind of alienation caused by the creation of a separate manager class.

A week later, in the same café, our co-author met with Ivo's son, who also worked at Uljanik. He describes how the price for certain services, which had been outsourced to subcontractors – ship painting, for instance – had doubled a few years back. When he asked at the shipyard why Uljanik was allowing itself to be milked like a cow, his line managers ignored him. Despite his criticism of these kinds of practices, Ivo's son generally sees the transformation in a positive light, though he does point out that crony capitalism and nepotism are widespread in Croatia. Not all the workers see the management as the only ones to blame for the shipyard's demise, however. Some workers describe how operations appeared frozen as if they were still in the 1970s. A young shipbuilder clarifies that the problem is less due to outdated technology and more because of "a lack of interest in new technology," although he puts this lack of interest down to the "workers' mindset" – older team members sitting out the time till retirement, doing no more than the bare minimum. He describes this attitude in one sentence: "Someone else will fix it." This statement also captures Uljanik's problem of a poorly developed sense of individual accountability and responsibility. As we can see, the systemic lack of responsibility has thus found its way into the vernacular narrative.

This criticism is not new, however. In an interview with a company newspaper in 1989, for instance, an engineer complained about the inadequate skills of the new workers "because the colleges don't produce the same [kind of] cadre they used to."[34] Even back then, another worker told us that one of Uljanik's most irritating problems was the fact that no one felt responsible for anything, something he blamed largely on the company management and the bloated administration.[35] The workers complained that once the reward system had been removed when socialism came to an end, special achievements were no longer recognized (the trade union made a similar complaint about the lack of wage differentiation according to individual effort). They lamented the fact that those who pulled their weight less earned the same as those who were more performance-driven. There was also a lack of efficiency and discipline. In other words, the workers were not exactly in favour of equal treatment for all. The grievances within the company could easily tip over into more general criticism of the system, much like the situation Michael Burawoy observed in his field studies in a Hungarian steelworks in the 1980s:

> Yet all around workers see inefficiency, injustice and inequality. Workers turn the ruling ideology against the rulers, demanding they realize the claims of their socialist propaganda. The state socialist bureaucratic regime of production sows the seeds of dissent rather than consent.[36]

The workers continued to take a critical view of the situation even after the system change, condemning the aforementioned crony capitalism and the broken promises of successive governments. However, during the interview with Ivan and his wife Jadranka, an interesting discrepancy comes to light. Ivan spent forty years working as a welder in Pula; his skin is pockmarked with tiny scars from the flying sparks he was constantly exposed to. He is nostalgic for the socialist Uljanik, remembering it as a much better system because "it looked after the workers, whereas the new system only looks after the managers." The latter know nothing about shipbuilding work. They are certainly no "experts" – they could not do the jobs of the workers under them and so never notice mistakes. He is convinced that it is this new system that destroyed the industry. His wife, however, has a different view. She believes the problem in Croatia is "wild capitalism," compared with "true capitalism[, which] is fair and just" – this is why, for instance, even the lowest paid workers in legal employment in Germany are able to live off their wages. The couple might have different views on Croatia's post-socialist period, but their perception of the "systematic destruction of the industry" reconciles their different perspectives – only their opinions on who is to blame differ.[37]

Something else that frequently came to the fore in our interviews in Gdynia were the critical views on the new workplace reality under capitalism, criticism that was rooted in socialist experiences and the hopes that were both dashed and fulfilled at that time. One of our interlocutors, G.K., introduced above, found a job in the equipment department of the Paris Commune Shipyard after graduating from the company vocational college in 1990. One year later, the shipyard changed its name and legal status and then, a few years after that, also its ownership structure – but this barely had any impact on the nature of his work. Nor, as our interlocutor explained, did it really change the atmosphere in the shipyard, at least in retrospect (the interview was recorded in 2016 and was one of the first to be conducted). He saw the shipyard's last twenty years, until its collapse in 2009, as consistent and coherent – despite all the upheaval. His sense of belonging to the shipyard community is a crucial factor in this coherence.

He juxtaposes this impression with his experiences at Crist, a company that moved to the former shipyard grounds after 2009. Crist does not build ships but rather different types of steel structures, including ship parts, largely ordered by West European customers. One of these structures – a large "steel crate" built for residential use and destined for an oil rig in the Baltic Sea off Denmark – was nearing completion at the time of the interview. Our interlocutor, who heads a team that performs finishing operations, invited us to visit his workplace. Finding a date for this unofficial visit during the final phase of such an important project proved quite difficult. However, after several attempts, on a summer morning in 2016, one of our authors was finally able to access the company premises and the shop floor of the former Stocznia Gdynia.

One thing that immediately caught the attention of the author was the massive steel structure, reminiscent of a multi-storey apartment building, which was standing in the "little" dry dock, ready to embark on the journey to its destination out in the Baltic the very next day. But what was even more interesting than these gigantic steel objects was the short interactions between our interlocutor and his workers and superiors. With the former, he used a mix of Russian, Polish, and Ukrainian to give the final instructions. His line manager, however, was a Norwegian engineer who was overseeing the project on behalf of the investor. The Norwegian asked questions in heavily accented English. It is presumably pure coincidence – but nevertheless relevant and enlightening – that almost all the workers we met at the shipyard (with the exception of our interlocutor) were not from Poland and spoke very little or no Polish at all. At most, the Polish language served to facilitate communication between the "Western" customers and the "Eastern" workers. This observation, along with the admittedly impressive but ultimately very fragmentary and "meaningless" product of the work of this international collective, provides a brief insight into the logic behind global capitalism in the context of European shipbuilding. A second insight can be gleaned from the fact that Crist uses the infrastructure of the erstwhile Paris Commune Shipyard for its production, yet fails to acknowledge this continuity.

Something else that became clearer during this visit was the difference between the work in the old Stocznia Gdynia and the new assembly shipyard. Although their functions are almost identical and are performed in the same place, sometimes even using the same machinery and equipment, the atmosphere described by our interlocutors and their colleagues was completely different. They portray a situation where the workers are detached, not like the big family they once were, a climate charged with feelings of suspicion and distrust, their roles task-based and completely subordinate to the new dynamics of transnational shipbuilding. This breaks with the principle of holism, which characterized work under socialism and was gradually dismantled during the post-socialist era, but which is still held in high esteem.

This caesura has different dimensions, the material being the most striking. In the past, the Polish shipyards built entire ships themselves, including equipment and machinery, from the keel laying to the launch. Today, however, with a few exceptions, the shipyards are chiefly suppliers. In the Polish shipbuilding industry today, a new production regime dominates where subcontractors or workers acting as "one-person enterprises" bear the brunt of the work – whether for complex technical design services or simple tasks.[38] This trend is accompanied by a change in the product portfolio. Most orders no longer entail shipbuilding in the strict sense, but rather the construction of marine structures – including the block of steel described above. Certainly, our interlocutor shows us the product with a clear sense of satisfaction that his company has managed to complete such a complex project on schedule – and emphasizes that this was in no small part thanks to his own personal commitment. However, compared with the collective pride that just a few years earlier was felt every time one of the Gdynia "giants" was launched, this is a very modest and personal sense of satisfaction. It is for good reason that there is no naming ceremony for a block of steel. The Uljanik shipyard in Pula, in comparison, operated as a "holistic" company to the very end, with all the well-known negative effects this had on its balance sheet. Here, in place of the holistic production that was maintained for far too long, there is now a void. Or at least no viable shipbuilding companies integrated into transnational supply chains have emerged in Pula so far, with the city and the region focusing entirely on tourism instead.

While the change in the semantic dimension of shipbuilding will be discussed in chapter 5, here we will address the collective side of the transformation. Building a ship together and then launching the vessel created a sense of communal esteem. Interdependent work tasks and an integrated production process meant the workers felt connected. This was a system where everyone had an important place; the work of one person was a necessary input for the work of the next, and vice versa. The impressive and "beautiful" ships, as many of the shipbuilders in Gdynia would describe them, gave rise to high-profile ship naming ceremonies. Even more important than the ceremonies, however, was the community that celebrated them (see figures 4.3, 5.2, and 5.3). This was not a sacral community, but rather a close connection between workers, mechanics, and engineers that grew from the bottom up on an everyday level. It was the connection of individuals through a shared mission, bringing together the problem-solving capabilities of thousands of workers in a complex, albeit never entirely frictionless, process that transformed nondescript raw materials into massive ships. All this created a special, affectively charged collectivity in the production halls and on the docks. The material product of the labour and the values of communality and solidarity it produced were inextricably linked to one another.[39]

Under socialism, these worker communities were further strengthened from above. The management organized celebrations and festivities, besides presenting awards, creating a social infrastructure, and reinforcing the role of the "collective." As reviled as the party-state in socialist Poland often was, it never tired of singing the praises of the working class and publicly giving it official recognition. When socialism came to an end, such references to the hallowed nature of the "collectivity" were ridiculed by neoliberal reformers and dismissed as pure propaganda. Yet they remain deeply embedded in the workers' minds. As they shared their stories, the people we interviewed articulated an affinity with the lifeworld of the shipyard. While this was rooted in the work carried out in the production halls, it also incorporated the social domains beyond the factory gates and the semantics of shipbuilding

Figure 4.3. Uljanik's electricians and their cleaning lady, late 1970s
Source: Private collection.

into one coherent world of experience. And while these feelings had already begun to dissipate in 1980, important elements survived until the closure of the shipyards in Gdynia in 2009 and Pula in 2018.

This bottom-up dimension of the lost shipbuilding community comes to light in different ways in our interviews. Our younger interlocutors, who, after the closure of Stocznia Gdynia, found jobs with different shipbuilding companies, understandably draw comparisons between the world of work then and now. The sense that the workers had lost social cohesion but also the importance of the work they once carried out is a recurring theme in these narratives – a sentiment that social anthropologist Don Kalb also encountered almost 20 years ago in his study of workers in Wrocław.[40] One of the people we interviewed is a former skilled worker from Stocznia Gdynia, who in the last few years, has worked for different (German, Dutch, and Swedish) West European companies. Our interview took place in a café chosen by our interview partner; he arrives in an expensive German car, one of the newest models, and at the end of the interview rushes off to his summer house in the Masurian lake district. This standard of living is one he could probably never have had, working for Stocznia Gdynia, yet he still assures us he would go back to his old company at the drop of a hat (D.P.).

Whether or not our interlocutor would really give up his new, well-paid job is open to question, but we see no reason to doubt the authenticity of his emotions – his yearning for a shipbuilding community. Such feelings are commonplace among our interview partners in Gdynia (and Pula), not only among the ordinary workers but also the engineers who, after 2009, found good jobs with shipbuilding companies on the Polish coast. Another skilled worker we spoke to told us that he now earns more money for less work but still misses the solidarity he remembers from his time at Stocznia Gdynia in the early 1990s (A.M.). He seems sure that his former colleagues feel the same.

In light of these statements from workers who have not been with the shipyard for very long, it is hardly surprising that our old communicants are even more nostalgic when reminiscing about their work for the former Paris Commune Shipyard (and this holds just as true for Pula). Not only did the shipyard define and give structure to the working day, but it also played a pivotal role throughout these workers' adult lives, from vocational college to retirement half a century later. Over the decades, the shipyard provided an opportunity, indeed almost a guarantee, for a stable and coherent life story. Anyone who worked for the shipyard would spend their entire working life, and in fact a large part of their private life, within the shipyard community. In the context of these kinds of narratives, privatization, downsizing, restructuring, and ultimately bankruptcy are not just peripheral phenomena, but are seen as ruptures in people's personal narratives, forms of expropriation, ruptures in the continuity of their identity.

An important means of creating this coherent experience of work was the training and qualification measures mentioned above. The most experienced workers we interviewed told us how older colleagues acquainted them with the work and craft of shipbuilding. Then, as managers, engineers, master crafts people, and team leaders, in time, they began to take on the very same role, guiding and instructing newcomers. The relationship between the senior workers and novices described by our interviewees went beyond strictly professional boundaries, manifesting an almost fatherly or paternalistic dimension. The image of a shipyard family that we encountered in so many of our interviews is clearly portrayed in these accounts. The use of familial and kinship metaphors is also a way of demanding recognition for the fact that the workers had "strong" emotional bonds that transcended the purely practical cooperation in the workplace. They implied a perceived right to protection from a "parent" and the expectation of reciprocity, as our hairdresser in Pula articulated. The workers were nestled in a community (albeit not necessarily the same one) and as such did not define their subjectivity as that of a monadic individual.[41] During the socialist era, this feeling materialized in frequent alliances between the shipyards' directors and workers, united in defence against the unreasonable demands made by "high politics" or in a bid to secure subsidies from the state. For the erstwhile employees and their descendants, the memory of the workplace community is firmly anchored in the abundant photos of the "collective" taken by the company photographers back in the day.

In this context, a moral economy becomes evident that is pre-transformational in the Polanyian sense: labour and economic activity were integrated into society and, in the relevant narratives, were underpinned by values of solidarity, mutual assistance, protection, and community. Many of our interlocutors idealized the strong ties, despite the fact that in those days internal clientelism gave rise to a lot of jealousy and inequality (although the workers' criticism of this aspect specifically showed just how important the social ideal of equal treatment was to them).[42] And the gender equality postulated by the regime had certainly not been achieved either. Nevertheless, the prevailing paternalistic care and collegial solidarity certainly saved the careers of some. One of our interlocutors, a foreman from Gdynia, recalls,

> Once there was this guy [a worker in my unit] who acted as if he was heading off to work every morning; he would pack his lunch, the sandwiches his wife had made for him, but instead he would go into the forest and sunbathe. I covered for him for three days. I protected him because he had two small children.... My goodness, if I had fired him, I would have his wife and children on my conscience because his wife didn't work. I went to visit him at home and asked his wife, "What's going on?" – "He's at work." – "What do you mean at work?" At the end of the day, he hadn't shown his face at work for three days.... So I waited for him. He didn't know I was there. Then, when he got home, I asked, "What's going on with you? If you don't come to work tomorrow.... And you'll be working double: instead of eight hours, you'll work 16, as punishment for the days you've missed. I didn't give up. I protected you from being let go." ... He came, and he worked those hours.... He pushed through and put it right. After all, he had his wife to think about. (C.Z.)

This incident is reminiscent of a popular subject of socialist realism literature, where the protagonists typically make mistakes before ultimately pulling themselves together and showing outstanding commitment. In the process they develop a high level of work discipline, thus becoming a happier and better person. But something this quote also highlights is a strong sense of almost family-like community, combined with a commitment to defending the model of the male breadwinner. From these memories, we can conclude that the communist ambition to create working class communities was, in some respects, successful – albeit only on the level of the company, or indeed the production hall, and not on the level of the much-vaunted "working class."

These structures and practices were particularly important for a city like Gdynia, whose residents in the 1950s and 1960s were mostly new arrivals. This was also reflected in the structure of the shipyard's workforce – the majority of workers came from all over Poland but many also arrived from the territories that had been lost to the Soviet Union. Most of the new workers were from rural areas and had no industrial experience to speak of. Given this rural conservative background, it is hardly surprising that it was mainly the men who went to work at the shipyard, while the women typically (or indeed exclusively) stayed at home as housewives. This was not least because the communist regime in Poland had been less consistent than other socialist countries when it came to promoting women's employment, especially in heavy industry.[43] Either way, the offer of housing was a particularly attractive incentive to work for the shipyard because the devastation caused by the Second World War and the postwar baby boom meant that living space was in particularly short supply in Poland.[44]

Pula, too, was a city with a new population. In 1946–7, it lost the majority of its ethnic Italian inhabitants as it became apparent that it would become part of communist Yugoslavia.[45] Many of the employees who started working for Uljanik's rapidly growing shipyard in the 1950s came from the countryside or other areas of Croatia and Yugoslavia. Nevertheless, owing to the shared experience of work and dedicated efforts by the company to integrate the workforce, a sense of community and solidarity rapidly developed. Large companies were thus an important site of socialization and stabilization in postwar societies, which were literally in constant motion, the trauma of the war still very much present among the people.

However, the aims of the communist regimes went beyond this. They wanted such factory communities to become a basis of unwavering support for their rule. In contrast to the other socialist states, the Yugoslav government sought to institutionalize the factory communities through the system of self-management. The idea was that employees would be loyal to the companies for which they worked and not to the country's different nationalities – the oft-invoked "fraternity and unity" was to emerge from this very shared experience of work and self-management. Poland's communist regime ultimately had to learn the lesson that one side effect of the strong community bonds they were so keen to develop within the company was the creation of Solidarność. The trade union movement was an abstraction of innumerable "small" solidarities on the level of the factory floor. When, starting in the 1980s, the shipbuilders began to bring their religious symbols and practices to the workplace, this showed yet again just how much oppositional potential the solidarity in the production halls held. The workers' ostentatious Catholicism provided them with a symbolic resource for their opposition to the state. And, at the same time, they demanded an even more holistic lifeworld. Contrary to the plans of those in power, the ships thus also produced dissent; yet, at the same time, they laid the foundation for the post-1989 memories that mourned the loss of this microcosm of socialism.[46]

This sense of loss is also widespread throughout the abandoned production sites in the original EU member states, albeit no longer as prominent because the closure of the shipyards in most cases was longer ago or the social basis of these memories has been displaced due to the gentrification of the dock areas. Among the workers in Gdynia and Pula, however, the longing for these bonds (as well as conditions) of solidarity and camaraderie still very much exists. Although this longing tends to be quiet and private, it still serves to build emotional bridges between the many people who continue to identify with shipbuilding. Memories of Solidarność, in contrast, are quite different – in today's Poland the movement is politically contested, making it a rather unfeasible basis for genuine acts of solidarity.

Fragile Communities: A Metaphorical Approximation

This account starts in the late-socialist period, an era when the Uljanik shipyard in Pula and the Paris Commune yard in Gdynia were directly and very actively involved in the lives of their workers, both during and outside working hours – and both behind and beyond the factory gates. To facilitate this, the shipyards created a community that was part of people's daily lives, a community that was celebrated, idealized, and staged, masking the actually existing inequalities in the company (such as between the white-collar and blue-collar employees). Party rhetoric and individual experiences of work as well as the welfare and social services and cultural activities organized by the shipyards were the backbone of this very distinctive community. Although this complex set of relationships and practices could not offset the structural deficits of the socialist system (including the inadequate supply of goods to meet people's daily needs), shipbuilders still remember it as the glue that held their lifeworlds together. When they reminisce, they talk about "the good old days," referring to the socialist version of their professions (which did not immediately disappear in 1989). They associate the socialist system of work with far more conviviality and solidarity than today's – albeit without harbouring any particular sympathy for the socialist system as a whole, and even less for the communist authorities. In keeping with the findings of historian Nina Vodopivec in her study of textile workers in Slovenia, our interlocutors' memories of socialism are not so much focused on the political system but more on a collection of workplace-related practices and relationships.[47] Their memories of the tensions within the shipyard have seemingly faded – even though articles in the company newspaper from the 1980s clearly attest to the existence of such friction.

The narratives we present here describe the post-socialist transformation years as a long and protracted process that ended this unity of experience. Not only did we ask our interlocutors about organizational change, but we also inquired about how they experienced it. In response they told us how corporate welfare was gradually dismantled, but they were much more effusive about the weakening of community ties. This second form of erosion is barely evident in the archival sources, which makes it all the more important for scholars to work with ethnographic methods. There are gradual shifts as well as continuities, but it is quite plain to see that the retreat of the shipyards from their role as multifunctional organizers was something people experienced as a loss.

What metaphors could be used to express the change in the community dimension of the shipyard described above? It is clear that neither the socialist nor the post-socialist company is a "total institution" along the lines of Erving Goffman's concept.[48] Its boundaries were too porous, the internal and external world overlapped in a volatile manner, meaning the company could not function as a sovereign institution in accordance with its own self-contained logic. Nor could the shipyards be seen as heterotopias in the Foucauldian sense.[49] For this, they were too closely connected to everyday routines and were not situated – either socially or spatially – outside the cities. They served less as a mirror reflecting social structures than points of intersection for those structures.

Would it make sense, in searching for an apt analogy for the nature of the shipyard, to look to the marine life that surrounded the docks? Perhaps the two shipyards conjure up the image of an octopus (or a leviathan)? Comparing the company with an octopus, we imagine its tentacles attaching themselves firmly to multiple spheres of life, controlling, and, when necessary, intertwining them at will. Yet, because of its slightly conspiratorial undertones, the octopus metaphor is problematic; in Croatia and Serbia, the same image tends to be used

Figure 4.4. Ship launch ceremony in Gdynia, 1970s
Source: Courtesy of European Solidarity Centre, © Janusz Uklejewski.

to describe the state as a corrupt institution.[50] Some of our interlocutors summoned up the image of an "anthill" as an emic metaphor for Uljanik, while in Gdynia people spoke of a "termite nest." It is presumably no coincidence that it is the workers that use these metaphors that reflect their agency and portray the shipyard's employees as bustling actors. Moreover, both images describe an apparently chaotic, but in fact strongly goal-oriented and united, collective undertaking. In the case of the shipyards, this materialized in the production of ships as well as in the organization of ceremonial launches (in Croatian *porinuće*, in Polish *wodowania*) and ship naming ceremonies, which clearly demonstrated the performative nature of communality (figure 4.4).

In the shipyards' own rhetoric, organic metaphors can be found in abundance. In a commemorative volume published in 2006 by the Uljanik shipyard, for instance, the island (*otok*) that is part of the shipyard is described as the "mechanical heart whose industrious giant cranes are proof of the shipyard's vital operations."[51] The big dry dock in Gdynia and the huge gantry crane towering above are also frequently referred to as the "heart" of the shipyard. All these naturalistic images, including the anthill and the termite nest, situate an important part of the shipyard in a wider, collective whole. They evoke the idea of an organism or body. All of the emic metaphors described express the notion of an individual that is part of a larger project.

It seems rather ironic that the communist regime failed to achieve its overarching ambition – to create a huge monolithic political body that places the common good above the interests of individual members or aspects of society – yet such feelings were widespread at the level of the company. Despite the changes in people's lives, these memories of erstwhile community and solidarity, as idealized as they may seem, create a new present-day community. This is a mnemonic community that takes shape as people tell their stories of the "good old days" or even in the course of an interview with a prying researcher. "Socially constructed images of the past unite individuals," writes Nina Vodopivec in her oral history of female textile workers in post-socialist Slovenia.[52] These memories, stories, and accounts make the discontinuity in people's personal narratives brought about by privatization all the more evident.

The justification for privatization was often that it would make employees co-owners, giving them a "personal stake in the successes and failures of the company."[53] The idea was for them to see the company's profitability as being in their own interests and for them to form a new community – this time of shareholders. And this very aim was what the management of Uljanik had in mind when, in 2012, they made the shipyard's employees the largest group of shareholders. Yet, our research shows that this move had exactly the opposite effect. Before privatization, the workers were willing to do overtime and showed a real dedication to solving the company's problems because they identified so closely with the shipyard. After privatization, in contrast, they began to detach their sense of self from the shipyard. However, both of these forms – before and after privatization – express the agency of the workers, who clearly did not simply passively accept the decisions of the political class and the management, but actively influenced them.

In this chapter, we outlined the process of erosion in a holistic industrial lifeworld – an experience that is manifested in subjective recollections of loss. The shipyards illustrate the demise of the Fordist model of industrial modernity (in its socialist form), where the factory is the main arena of the workers' professional and social life as well as the locus for their self-realization as citizens.[54] More than half a century ago, this process of decline began as structural change in the "West" and was exacerbated as Western economies were forced to adapt to the new conditions after the oil crisis.⁵⁵ As shown, it was another fifteen years before the structural change reached the heavy industries of Eastern Europe, arriving as communist rule came to an end, in the form of a transformation that was teleologically conceived and yet ended up being chaotic at the same time. Although nowadays, the "transition from the planned to the market economy" can largely be seen as being complete, the case studies presented here suggest that the holistic model of industrial life continued to exist for far longer than many of the apostles of neoliberalism had predicted – and, as an ideal, never in fact disappeared. When looking for differences between West European structural change and East European transformation, this experiential dimension can, to some extent, help us understand them.

Over the years, post-socialist companies became less "social" and more "professional," concentrating on performing specialized tasks, generating profits, and heightening disciplinary and safety measures (or they simply went bust). The managers and political leaders called on the individuals to make sacrifices – no longer in the name of the community and solidarity, but for the sake of efficiency and productivity. In her study *Privatizing Poland*, Elizabeth Dunn comes to the following conclusion:

> By stripping workers of their social context, devaluing their personal connections, making their family relations irrelevant, and dismissing their moral beliefs about interpersonal obligations, new management technologies attempt to make Poles into the market-rational invisible subjects of post-Fordist neoliberalism.⁵⁶

From a historical perspective, the results of this ideological project seem ambivalent, at least with respect to our two case studies. It was not only the fact that shipbuilding was a skilled trade or the dynamics of the industry but also the many "imperfections of the market" during the process of transformation in Croatia and Poland that lent business practices and employment relations a strong interpersonal dimension. This was rooted, on the one hand, in the inertia of the local lifeworlds and, on the other, in the specifics of the capitalist system in post-socialist countries. Croatia's notorious cronyism, in particular, as well as the close ties between the shipyards and the governments in both countries, enabled the continuation of community-oriented practices, which ought not to have existed in a neoliberal context. Thus, as already described, Uljanik continued to employ talented footballers so they could play for the company team and helped repair the roof of the municipal library, going above and beyond the paradigm of "social corporate responsibility,"⁵⁷ to name but two examples. In addition, staff still socialized with each other outside work and would sit together in the *marenda* during their lunch break. There was no sudden "shock therapy" or mass layoffs. Instead, the workforce was gradually cut back, while the company tried to continue "business as usual." Much like the transformation at company level, the changes to the foundations of the community and of the social dimension of human existence, too, were gradual, nonlinear, and marked by ambiguity.

But we should not be looking at this story from its conclusion. The workers we spoke to did not do this either, instead keeping a holistic view on their own life stories. This is linked to a form of holistic production that continued to be practised long after the systemic change of 1989/91. The Uljanik shipyard and Stocznia Gdynia continued to manufacture a large number of their components themselves (including engines and equipment): they created their own blueprints, trained their workers in company-owned vocational colleges, and provided budding engineers and business administrators with scholarships to study at local universities – despite the fact that none of this made them a profit. They almost exclusively employed local workers and presented their products as a symbol of "Polish" or "Croatian" shipbuilding. This business strategy – which no longer really belonged in an era of transnational supply chains with highly specialized businesses – was also an important reason for overproduction.

This holism of production, community, and way of life could survive in a capitalist environment only provided that the national and local government continued to provide sufficient subsidies. And while this financial aid may not have been enough to preserve the holistic system in its entirety, it was sufficient to prevent it from disintegrating immediately. With the helping hand of the state, the shipbuilding octopod was able to change its colours and adapt to the new environment – at least on the surface. After EU accession, however, this mere change of form rather than substance was not enough to make an impression on the Eurocrats in Brussels.[58] They did not believe shipbuilding was future-proof, unlike the aircraft industry, for instance. Thus, one of the reasons our two shipyards and their West European companions in misfortune were unable to ascend in global capitalism was because they had so much social traction holding them down. This was no longer the order of the day, which is why the old (post-)socialist tankers of industry suffered the same fate as their erstwhile competitors in Bremerhaven, Dunkirk, Glasgow, or Liverpool – all sinking like stones. But can shipyards still create meaning, even if they have stopped building ships? Can people make their industrial environment "habitable," even if they can no longer rely on a secure income? These questions are just as pertinent for Eastern Europe as for the West.

5 Added Value: Ships, Labour, and the Production of Meaning

In his track "The Last Ship," Sting sings about "the roar of the chains and the cracking of timbers" as "a mountain of steel makes its way to the sea." In this song, from the album of the same name, the world famous musician is singing about his hometown Wallsend in the north of England, which, for 150 years, was home to one of the biggest shipyard in Britain and thus also the world. Both the father and grandfather of the singer-songwriter worked for the shipyard – as did the majority of the men living in Wallsend at that time. In 2007, long after its boom years had ended, the shipyard was finally shut down. With this song and the eponymous Broadway musical, Sting is paying tribute to the world of shipbuilding in which he grew up – an empathetic commemoration of the shipbuilders and their culture. At the same time, the song also contains a universal message as it talks about bidding farewell to a specific form of industrial modernity in days when heavy industry went hand in hand with the production of social relations and cultural meanings.

In both the capitalist West and the socialist East, shipyards were emblematic sites of industrial modernity. They were associated with wide-ranging ideas of progress, prosperity, and secure jobs. It is important to make clear at this juncture that the notion of West and East is a binary juxtaposition that fails to adequately capture a multitude of similarities between the two "blocs" with regard to industrial organization and employment relations.[1] Work in shipyards, in East and West alike, was seen as hard but worthwhile, with the employees enjoying a not inconsiderable level of autonomy in the production process as well as relatively high wages. They took pride in the fruits of their labour and benefited from the formal and informal welfare services that large industrial firms provided for their workforce at the height of Fordism. The same applied to other mainstays of industrial high modernity, especially the steel, coal, and automobile industries. Although not the only branch to do so, heavy industry enjoyed a prominent position in the public perception of economic and social life. And this was something the factory workers in both the West and the East were able to benefit from during the postwar boom decades. The communist ideologues and the increasingly powerful Western European social democrats alike lauded the "ordinary" workers' contribution to the postwar reconstruction, growing prosperity, and overall development. Even right-wing dictatorships devoted themselves to "development" by means of industrialization. Among them were Taiwan and South Korea, where, incidentally, shipbuilding played a pivotal role in industrial policy, or 1960s Spain and Portugal, whose government propaganda attached special importance to industrial labour.

Until the start of structural change, the differences between the shipbuilding industries of Eastern and Western Europe were negligible, at least on the shop floor level. That being said, up until 1989, the Eastern Bloc and Yugoslavia had been shielded from the mass job losses and the shift towards the service economy, whereas in the West, shipbuilding work began to change dramatically in as early as the mid-1970s. When this change did eventually arrive in Eastern Europe as well, the transformation was all the more rapid and unsettling. Owing to its sheer pace and the different political and socio-economic conditions and impacts, structural change in the East took on an entirely different quality to that in Western Europe. And its is these differences we will now trace, primarily in the workers' daily lives and worlds of meaning as well as the significance they attached to their work and the products of their labour.

The size of a ship is usually measured by its displacement tonnage or its load-bearing capacity. The shipyard workers saw the total gross tonnage of the ships produced in a given year and the volume of orders – rather than annual profits – as the indicators of their companies' success – or later failure. Yet the ships produced in Gdynia and Pula were even heavier than the steel used to build them: every tonne carried the additional symbolic weight of the ships' importance. The vessels were laden with exertion, hope, love, and, most important of all, pride (with the occasional measure of reluctance and exhaustion thrown in). The ships' interiors were filled with the multidimensional associations made by individual protagonists – the state, the shipyard, the city, the engineers, and last but by no means least, the workers themselves. The shipyard was thus so much more than just a place where ships were made; it was also a complex apparatus of the production of meaning. The ships were literally overdetermined with meaning, overloaded to such an extent that the profit dimension was often seen as downright banal.

The cluster of shipbuilders along Croatia's Adriatic coast and the Polish Baltic Sea should be seen as representing for their respective countries "what the cities of Turin or Detroit represent for the history of the working class in Italy and the USA"[2] – or indeed what the Wallsend that Sting sang of represented for the north of England's working class. Even though Pula and Gdynia did not inspire a melancholy song by a world-famous popstar, here, too, the workers' worlds of meaning were shaken up by the fundamental political and economic changes that took place after the 1980s, particularly once communist rule came to an end. Although "1989" cannot be seen as an absolute turning point (in Yugoslavia, like in the Soviet Union, the decisive political watershed did not come until 1991 anyway), it nevertheless accelerated the change in the symbolic site of industrial labour in the social imagination.

While under state socialism, public and personal attributions of meaning to the (heavy) industrial workplace were largely congruent – we need only to think of the figure of the heroic worker here – after 1989, this consensus began to wane. In political and media discourse, industrial labour lost its privileged position, those in power no longer saw the benefit of singing its praises, and in some post-communist contexts, it was outright reviled as an anachronism.[3] The symbolic charge of labour shifted from contemporary public discourse to the world of personal but also of shared memories. In what was once state socialist Eastern Europe, the "post-coal and steel"[4] era began. Although this was later than in Western Europe, the change in the political and symbolic site of heavy industrial labour was all the more sudden and rapid. The specific heuristic valence of post-socialist deindustrialization reminds us instantly "of the actual catastrophes of industrial labour and the particular values and social relationships that were the by-product of the factory."[5]

The reason shipbuilding lost its economic importance and poetic charm in the eyes of the political decision-makers was largely down to external forces, or rather actors – specifically the global market and European Union – forcing a change in mindset. With the shipbuilding's symbolic overdetermination fading away, the governments and increasingly the public, too, began to lose patience with this industry and its constant loss making. Many workers sublimated their emotional connection to their jobs and the product of their labour into a nostalgic memory of what seemed like better times. And this is also where the future-oriented association with progress came to an end. On a personal level, much like within the companies themselves, it was no longer a matter of advancing or growing, but rather just muddling through.[6] The sense of belonging to a community through a connection provided by one's work went from being a social reality to no more than a memory. In this context, ships were of special symbolic significance; after all they are the biggest movable objects that can be built by humans and allowed each and every worker to take some of the credit for this achievement.

The Reduced Value of Industrial Labour: From Communism to Post-Socialism

The assertion that industrial workers, especially in heavy industry, played a key role in the self-perception of the communist regime needs no real explanation. Even though the links between the industrial workforce and government power were at best ambiguous and at times controversial and conflict ridden, as the countless workers' protests in the Eastern Bloc and Yugoslavia attest, the communist regime continued to rely heavily on the close relationship with the "working class" for its legitimacy.[7] It was rare for an illustrated volume about the achievements of communism or the splendour of the country not to feature photos of factory workers smelting steel, operating gigantic machines, or welding one thing or another. Welders – even the rare female one – were crucial for shipbuilding and thus became a particularly popular symbol of the socialist working class (see figures 5.1 and 5.4). Charity Scribner comments that "socialists aspired not only to lift the Genesis curse of labour, but rather to find collective and creative fulfilment *in work*."[8] The discursive upgrading of the industrial labour force was reflected above all in the welfare policies of the 1960s and 1970s, which sought to improve workers' living conditions. In a kind of "socialist paternalism," to use Katherine Verdery's phrase, the communist regime placed its promise of social redistribution at the heart of its ideology and public self-representation. For quite some time, this enabled it to achieve at least a certain degree of approval among the population.[9]

The "workerism," to use a term coined by Mark Pittaway to refer to the approach taken by the communist regime that had resulted from the upheavals of the 1950s,[10] did not, however, provide everyone with the same level of welfare and prosperity. The predominantly male heavy industrial workforce enjoyed better welfare services and higher wages than those working in light industry, which employed a significantly larger share of women. The communists were similarly unsuccessful in their attempts to close the growing divide between the East and the West when it came to the material standard of living. These experiences and how they were manifested in public and in private formed the backdrop against which the workers articulated the meaning that was also produced on the factory floor. Such experiences then became

Figure 5.1. Welding work being carried out on the hull of the *Bailadila* supertanker in Uljanik, 1971
Source: Courtesy of Historical and Maritime Museum of Istria.

memories that would reshape the workers' views on the past, the present, (and future) of industrial labour. The past cannot determine the future; but it is vital part of the framework that people use to order their experiences and make sense of them.

It is no secret that during the era of state socialism, a great many workers and other citizens throughout the communist world were dissatisfied with their working and living conditions – sometimes even taking the risk of participating in public protest. The wave of strikes carried out by Polish workers in 1970 and 1980, in each case sparked by poor living conditions and price increases, as well as the multiple walkouts in Yugoslavia, are prominent examples of the discontent among the "working class." Yet, many of the studies on the collective memory of industrial workers based on ethnographical observations and oral history show these workers' memories of the state socialist era to be overwhelmingly positive.[11] It is not our intention to go into too much depth on the phenomenon of "nostalgia" – suffice to say that it is widespread in post-socialist Europe and even those with anti-communist attitudes often perceive the transformation as a "demise." Studies covering quite different places and sectors, such as Chiara Bonfiglioli's work on female textile workers in the former Yugoslavia or the contribution by Jeremy Morris on a post-Soviet monotown focused on heavy industry, encounter a similar "before-and-after" narrative. In this narrative, the "before" (of socialism) is always seen as better, at least for workers, with the time that has passed since then being described as demise, disintegration, and expropriation.[12] Commenting on this, social anthropologist Don Kalb states, "People complain that what they got from socialism – the chance to make and sustain a family, build a career around honest work, and maintain a house of one's own – can't so easily be gained today."[13]

Such memories and sentiments point to the cogency of the European (and also North American) postwar model of an industrial society that promised factory workers secure jobs, an increasing standard of living, and a predictable path to joining the middle classes. This was also a time when big factories offered extensive welfare functions that provided an infrastructure for solidarity and sociability. As a result, oral history projects on de-industrialized towns and cities in the West come across very similar attitudes to those observed in post-socialist Eastern Europe.[14] Beyond the specific locational characteristics, general patterns relating to the experience of the loss of meaning and importance of heavy industrial labour are also evident – whether in Gdynia and Pula or Bremerhaven and Glasgow.

Although these narratives quite clearly do not depict the past in an unembellished and neutral manner, but rather filtered through the experience of transformation and of the given situation in each case, they must not be dismissed altogether. Instead we should ask ourselves why these positive memories dominate to such an extent after 1989 that they obscure the earlier complaints and criticism of the socialist system – even among active strike participants or those involved in Solidarność, the large-scale protest movement of 1980.[15] For us, these memories therefore have a heuristic value. On the one hand, they are an important instrument enabling individuals and groups to position themselves in the present, to make demands for the future, and to develop a coherent sense of self. As Tanja Petrović writes, "Remembering socialism as a historical memory – like any other memory and any other historical legacy – greatly shapes people's present and gives meaning to the social structures, values, beliefs, and actions of a society and its members."[16] On the other hand, these memories, albeit mediated and influenced by different levels of medialization, reflect what aspects of the past people cherish. The recollections presented in our case studies at first seem to confirm commonly known research findings. They show that the communist ideology's position on industrial labour influenced the workers' self- perception – despite the fact that they were perfectly well aware of the hypocrisy inherent in that ideology.

In an essay focused on remembrance culture, Serbian historian Predrag Marković emphasizes what he calls the "S" values: solidarity, security, stability, social inclusion, sociability, strength, and self-esteem (recognition).[17] These were values that the communist regime promised to uphold and did in fact implement, to some degree at least. Indeed, they also provide the context for the nostalgic accounts given today. Although, for each of these values, it is easy to identify a substantial gap between the political promise and lived reality, it is still undeniable that these are recurring tropes that workers use to give meaning and a sense of purpose to their often precarious situation. We are thus not seeing cognitive dissonance here, but rather subjectivities that were called into question by the fundamental transformation of lifeworlds and discursive context after 1989.

In the early 1990s, the industrial workers had to deal not only with a rapid decline in their standard of living but also with the sudden loss of the prestige that had been ascribed to them. What is more, these workers were no longer present or visible in the public sphere. Today, no PR manager in Istria would ever come up with the idea of including photos of factory workers in advertising material used to promote the region. Even the subsequent economic growth – which, in Poland, purely numerically speaking, was almost on a par with Germany's postwar economic miracle – failed to heal the wounds caused by the initial economic slump and social decline of the early years of "transition." According to one estimate, the total value of goods and services produced in the "economies in transition" fell by as much as a quarter from 1989 to 1997.[18] It was not only the social and symbolic status of industry and industrial workers that was lost in this process, but all hope that industrial modernity meant the first step towards a utopian future – a belief that was maintained for much longer under state socialism than capitalism.[19] In many respects, the crisis of the 1990s played a crucial role in shaping today's memories.

In 2008, referring to Romania, social anthropologist David Kideckel came to the conclusion that "concern for workers' conditions is marginalized and delegitimized, and many branches of industry are scorned as socialist survivals."[20] Once highly valued, the workers were now denigrated. In the case of Romania – or so says Kideckel's hypothesis – this was based on a deliberate political strategy: the new elites saw the workers from the "old" industries as backward and as an obstacle on the path towards a developed capitalist society. The workers lost the social and symbolic capital they had been able to accrue under socialism, and with it their power for collective action. At the same time, the media began to portray the service professions, "flexibility," and entrepreneurship in a positive light. In a similar vein, in her study of the privatization of a Polish factory, Elizabeth Dunn observed that the new management was constructing the shop floor workers "as products of the socialist system, who are unable to adapt to changing economic conditions, precisely *because* they lack the ability to think." The advocates of the market economy not only deemed socialism unreformable; they also considered factory workers "untrainable and unchangeable."[21] Even renowned Polish reformer Adam Michnik ridiculed the "ex-socialist" workers, claiming in 1999 that their main job had been to produce busts of Lenin.[22] Workers were blamed for their own suffering. "Class" disappeared as both an analytical and political category, meaning it no longer existed as a group in whose name political demands could have been made.[23]

The demise of the proletariat was not merely a symbolic creation, but also the result of political reforms and the weakness of the trade unions, which had experienced a loss of both trust and members.[24] In Poland, this was perhaps more surprising than elsewhere; after all, this is a country in which a true workers' movement, Solidarność, had played a key role in weakening and ultimately destroying communist one-party rule. However, as expressed by the evocative title of David Ost's influential book, the victory of the movement after 1989 brought in its wake a "defeat."[25] On this, Scribner comments dryly, "The greatest event in Europe's postwar labour history contributed to the erasure of the movement's conditions of possibility."[26]

In the 1990s, when Solidarność was transformed from a social movement and trade union into a political party, it began to ignore the workers, concentrating mainly on pro-market reforms, a vigorous anti-communist policy, and – as part of the governing coalition from 1997 to 2000 – on the promotion of "Christian" values and anti-communist lustration, rather than workers' rights.[27] A second important political factor in the decline of the working class was the change in direction of the liberal intellectuals, who, having been on the side of the industrial workers in the 1970s and 1980s, now increasingly adopted pro-market positions. Their most important mouthpiece, *Gazeta Wyborcza*, a daily newspaper that had been officially affiliated with Solidarność up until the mid-1990s, now regularly portrayed the trade unions as vestiges of the past that were jeopardizing the necessary economic reforms.[28] Even Solidarność opposed trade union mobilization in the private sector out of fear that this would put off foreign investors. As a result, the level of trade union organization in the private sector was particularly low – in 2017, just 2 per cent of private sector employees in Poland belonged to a union, compared to 28 per cent for the public sector.[29]

The Croatian trade unions did not escape this development either. In the first decade after independence, the level of union membership in Croatia dropped from 90 per cent of employees to around 45 per cent and continued to fall steadily from then on, reaching around 23 per cent in 2016, which was substantially higher than in Poland but still showed a significant discrepancy between the public and private sector (68 compared to 17 per cent).[30] The strong fragmentation of the trade unions played a key role in this. Marina Kakanović emphasizes that concepts like "collective wage agreement" or "strike" – in other words, the language typical of the working class struggle – increasingly sounded "socialist-like" or even "anti-Croatian."[31] Under these conditions, the many small workers protests seen in Croatia were unable to merge to form a broader movement, which could have defended the social rights and symbolic position of the industrial workers. Initially during the course of nationalist mobilization and then, more vigorously, during the 1991–5 war – the image of the worker was rapidly forced out of the public sphere. President Tuđman even tried to replace the term *radnik* (worker) with the new name *djelatnik* (employee), which sounded more Croatian (the Serbian word for worker is also *radnik*) and had less of a working-class connotation. An even more impactful step was the replacement of the word "worker" with "Croat," as the governing HDZ party drew its legitimacy not from a claim to defend workers' rights but rather from its role in the Croatian process of independence and ultimately from successfully defending the country against Serbian aggression as well as liberating the Serb-occupied areas.[32]

The politically motivated degradation of the workers was further reinforced by the deep economic crisis. At the end of the 1990s, Croatia's GDP was lower than it had been at the beginning of the decade. Unemployment was at around 20 per cent – and if we include the workers who did not receive their wages, this figure was as high as 30 per cent.[33] In the end, Croatia transformed into a service economy and tourism became the country's most important industry. According to one Croatian sociologist, waitresses and waiters (*konobarica/konobar*) became an emblem of the new division of labour, while in the 1980s, industrial workers (including miners) were the country's largest group of employees.³⁴ In 2001, half of all those in gainful employment worked in the service sector (while the manufacturing industry still accounted for just over a quarter of total employment). Domagoj Mihaljević argues that certain political decisions were the main reasons behind the real and symbolic decline of industry – these included, for instance, the pegging of the national currency to the German mark and later the euro, which led to the overvaluation of the Croatian kuna, thus putting domestic export companies at a disadvantage. This resulted, after 2000, in dependence on foreign banks, whose loans primarily drove property construction and private consumption, but not investment in production capacities.³⁵ Society's ideas of what constituted a good life – in other words, the social imaginaries – had thus been fundamentally transformed, to a large extent through political intervention.

There were however some exceptions to the structural change that was seen across the region. The shipyards in Pula and Gdynia resisted rapid deindustrialization. The following pages will, however, focus less on the companies and much more on the worlds of meaning of those who held various positions in the shipyards. We mainly spoke to people who had spent the majority or even all of their working lives in the shipyards and thus, over time, had come to strongly identify with their work and the company. A look at the in-house newspapers confirms our findings for the late-socialist period. Our protagonists do not only interpret their work experiences and the importance of the jobs through the prism of their personal memories. The subsequent dismantling of their stable lifeworlds and the resulting fears and dashed hopes also had a significant impact. These memories are not representative in the sociological sense but they do provide insight into the symbolic importance of shipbuilding under socialism and the long shadows this cast in the period that followed. This shows that the memories of the romanticized old shipyard community were not just a nostalgic reflex but also a foundation that enabled people to cope with the here and now, that made it "habitable" despite all its contradictions, as social anthropologist Morris so aptly expressed in his study of a de-industrialized city in Russia.³⁶

The Work Ethic and Solidarity of the Past

"It was wonderful. We worked," reminisced one of the former shipyard workers we interviewed in Pula. Many of our interlocutors told us how "happy" they had been to have jobs in the shipyard. One expression of the emotional connection with work was the flowers, plants, and small trees that, in 1990, two workers from Pula bought from the market with their own money and planted on the grounds of the shipyard.[37] This connection was manifest not only in the context of the factory premises that were now home to the workers, but also in the distinct sense of responsibility. Also in 1990, long-serving welder Ivan Zenzerović told the company newspaper *Uljanik*, "You have to love your job." He went on to declare that every duty must be carried out "properly," with precision, responsibility, and care, then adding in more general terms, "Work makes people happy."[38]

This attitude reflects, on the one hand, the understanding that shipbuilding was very much a skilled trade that even after the Second World War continued to be more workshop oriented than other highly automated sectors. The skilled workers in particular were very much aware of the essential role they played in the production process. In 1981, more than 40 per cent of Uljanik's employees were qualified or highly qualified and more than 8 per cent had a university degree. A brochure targeting young people listed twenty-two different shipbuilding trades – all of which could be learned at Pula's shipyard and technical college.[39]

While these factors – at least historically – were also present in shipyards in countries beyond the socialist world,[40] under state socialism workers seem to have identified even more with their jobs: the workers felt they were given proper acknowledgment – not only by colleagues and superiors within their work organization, but also by the communist ideology as a whole. Tanja Petrović comes to a similar conclusion in her study in which she interviewed workers from a cable factory in central Serbia: "The past is always perceived as better, not only in material terms, but also in terms of dignity and the respect workers enjoyed under socialism."[41] It thus seems that industrial modernity not only alienated workers from their own labour and the fruits of that labour, but also saw employees identify with their work, provided they were granted a certain degree of autonomy within the production regime and their experiences concurred with the official narratives.

The workers' commitment to their jobs was expressed not only in words. One of our Polish interlocutors, who had worked as an engineer at Gdynia Shipyard, began the interview by showing us a framed photograph displayed prominently on the wall of his small apartment. The photo showed a massive ro-ro ship used to transport rolling stock that he had designed towards the end of his career. He proudly told us all the technical details. The fact that our interview partner was one of the organizers of the protests that took place in the shipyard in 1980 and went on to play a leading role in his company's Solidarność group does not detract from the sense of identity and belonging he felt with the shipyard, although he certainly makes no bones about his critical stance on communism (A.K.).

In Western post-industrial contexts, too, work ethic plays an important role in the workers' memories, as Andrea Muehlebach highlighted in the example of Sesto San Giovanni in Italy.[42] This would appear to be a powerful pattern uniting the cultures of industrial workers across time and space. A clear illustration of this can be found in the recollections of an erstwhile factory foreman we interviewed in Gdynia. "There was a time when I believed I was an irreplaceable part of the shipyard," the shipbuilder told us, going on to provide a passionate account of how they once came to pick him up from home because they had problems with an anchor chain. "The floor manager came to my home and told me: 'The car is downstairs. We can't fix the problem. You have to come!'" (C.Z.). Once he got there, our interlocutor was able to make full use of his experience. He recalled the incident:

> I just switched a torch on. Where is it caught? At this point here, and that point there.... "Cut that bit off for me," "Let's do a bit of welding here," "Drop the anchor!" "Pull the anchor up!" The noise has gone! ... I've done it! Then, in the middle of the day, they drove me home again. The managers were happy because I fixed the problem quickly and easily.... Because of all the years I'd worked there, I had a lot of experience. Experience makes all the difference.

This anecdote reflects the narrator's pride at how well he had mastered his trade. It also clearly illustrates the informal relations between the shipyard workers, which traversed the formal hierarchies and dependencies, including those between the blue-collar and white-collar workers. Ultimately it is a story of human agency, of resourcefulness, and creativity in a situation where the formalized production process had failed – as was all too often the case under state socialism.[43] Last but not least, this also makes it a story of structural problems (technical deficits, unrealistic plans, and inadequate quality controls), which were resolved thanks to the personal commitment and dedication of the workers or group actions. Thus, ironically, the weaknesses of socialist production created not only frustration but also a feeling of self-worth, a sense of "We did it!" – even in the face of adversity. In the accounts of our interlocutors, this "we," however, generally referred to a male collective, as there were far fewer women in the technical and manual occupations (see the photograph of Pula's electricians, figure 4.3). In this anecdote, all those involved automatically assumed that the wife of the foreman would stay at home to take care of the family and the celebratory Easter meal.

The working day at Uljanik started exactly as it did in Gdynia with the shipyard's workers passing through the factory gates or *kapija*. This marked the crossing of the physical boundary to the workplace, where the production regime made its voice, or rather its "siren," heard: "You should have seen how we'd all get up in the morning and run to work, wondering whether we would make it through the gate [in time]" (J.L.). Two sirens were sounded at the start of every shift; if work began at 6:00 a.m., the first siren would go off at 5:55 a.m. and the second at 6:00. But the gate would already be locked after the first siren because the workers needed at least five minutes to walk to their work stations. This was a profoundly formative experience, also from a moral perspective. For some workers, arriving late was simply "inconceivable," as, according to J.L.'s recollection of the 1960s, this implied dishonest, undisciplined behaviour overall. The work ethic was so deeply internalized that every time a worker was late, they would feel genuine "shame." Some workers even took a day off rather than show themselves up in front of their colleagues by turning up late (B.Bu.). For many, being a decent worker was unquestionably an integral component of meaningful work.

However, part of this distinctive work ethic was also a well-measured lack of discipline, while those who took things too far were looked down on by their colleagues. As Michael Burawoy outlined in his analysis, the workers enjoyed extensive autonomy in the socialist "factory regime."[44] According to A.K., erstwhile chair of the disciplinary committee in Uljanik, "it was a bit more difficult to impose discipline on the workers in the former Yugoslavia than it is today ..., because they were always right." There were all manner of rules but, at the end of the day, the workers had "more rights" than the head of the work organization, which made it "hard, very hard," to enforce anything.[45] Another worker reported that, because of the power of the trade union and the party's worker-friendly position, workers that were wont to slack off enjoyed a good deal of protection: "You couldn't fire a worker, definitely not, it was impossible. You had to be a real scoundrel to be out on your ear" (L.Ž.). This last comment concurs with accounts of people's experiences in other state socialist societies, where the shortage of workers, especially skilled workers, made the management reluctant to dismiss anyone for non-compliance with company discipline rules.[46]

That said, not all rule-breaking remained within the acceptable limits, which is why the management did not rely only on the workers' internal discipline. A detailed analysis of the records of Uljanik's disciplinary committee shows, for example, that workers who failed to show up for work repeatedly or for long periods, without permission, were in fact let go.[47] These breaches of the rules also included widespread theft, especially of construction materials, something that was viewed ambivalently, though. This type of theft was a phenomenon that the factory directors in all socialist countries were familiar with and one that illustrated the agency of the workers. In the socialist system it was also associated with a personalized "economy of favours." An additional element was the appropriation of that which, under the Yugoslav self-management system, officially belonged to the workers anyway. Practices like this came to an abrupt end in the early 1990s, when the new management imposed strict top-down discipline, seen by some of the workers as degrading and disempowering.

Beyond the regulations and sanctions, there was quite clearly still considerable leeway, with employees being free to domesticate the shipyard to their own ends. This is illustrated by an account from an employee of Uljanik in the 1980s:

> The *marenda* [lunch hour] was at 10:30 a.m., we stopped work at 10:00 or 10:15 at the latest, cleared away our tools, if necessary, depending on who worked where. And then we walked to the kitchen area, ate our lunch, had a drink – we made a symbolic contribution in payment for our *marenda*, a ridiculously small sum. For a while, we were also given half a litre of milk, then for a while after that we got two glasses, each with 200 millilitres, and so on. And then, after lunch, we worked another couple of hours, we did a little bit of work for three hours, dum dee dum. Those who didn't want to work overtime then went home, then about fifteen minutes or half an hour before we finished work, we went for a shower, got changed, then off we went home. At about 3:00 p.m. in the winter or 4:00 p.m. in the summer.

Experienced workers like the one we interviewed here knew which minor violations of the rules were acceptable – to their foremen, their co-workers, but to themselves as well, since their sense of dignity was built both on their strong work ethic but also on their refusal to allow themselves to be ordered around. For instance, it was the norm for the workers to finish at 11:00 a.m. on Fridays, then stay in the shipyard until the official shift end (3:00 p.m.), barbecue some meat and eat and drink together (Z.M.). Although alcohol consumption was officially forbidden, it was commonplace – the many sanctions meted out by the disciplinary commission are testimony to this. Some even sold homemade wine or *rakija* (schnaps) from a cask they kept in their lockers. Despite these transgressions, which were important in order for the workers to see (and present) themselves as shrewd and autonomous protagonists, overall, a strict and serious work ethic prevailed. This was predominantly down to the fact that the workers kept an eye on one another, keen to make sure no-one stepped out of line too much as well as the fact that, as the opinions expressed in the *Uljanik* company magazine in the late 1980s show, a certain spirit of competition clearly prevailed between the different departments. In the post-socialist world, all this was replaced solely by work discipline decreed from above.

A cornerstone of the socialist work ethos was the wages of the shipyard workers, which, compared to other industries, were relatively high. However, employees also had other motives for doing their work properly. According to Č.Č., "It wasn't just about the money." This perspective highlights one of the differences between then and now: today, financial remuneration is largely, if not exclusively, the only thing that matters. During the interviews, some of the workers emphasized the prestigious position occupied by the *uljanikovci* in Pula's social hierarchy. In the socialist era, they were seen as "top dogs" (Z.M.). Whenever anyone reported having landed a job in the shipyard, this was often received with the exclamation "Oh, you lucky thing!" (Č.Č.). This perception was also actively nurtured from above and supported by the shipyard management. In a speech to shipyard staff in 1980, director Karlo Radolović repeatedly referred to the "honour" and "reputation" that had to be defended.[48] He reiterated this in 1986, declaring, "Today, Uljanik represents our honour," going on to emphasize in the very next sentence that "Uljanik's greatest assets are its good and honest workers."[49] Local economic policymaker Vinko Jurcan, too, confirmed the social prestige of being employed by the shipyard: "Anyone who works for Uljanik should be very proud." Long periods of service were celebrated by the shipyard accordingly. For twenty-five years of service, for instance, workers would receive a certificate, and this would be followed by a toast during which the alcohol flowed in abundance (figures 5.2 and 5.3). The importance of such events for the workers is not to be underestimated – it was during these moments that they felt acknowledged and valued, and at the same time, were able to have a good old knees up with their colleagues, who were often also their closest friends. Rituals like this connected the life stories of individuals and groups with that of the company.

The unique self-esteem that shipyard workers felt as well as the social recognition they enjoyed were of course inextricably linked to the product of their labour – the ships. The documentary film *The Colossus of the Adriatic* (1972), about the giant tanker that was built in Uljanik in the early 1970s, features skilled workers and foremen as well as female workers, engineers, and members of the management. They tell of the

176   In the Storms of Transformation

Figure 5.2. Uljanik workers receiving certificates for twenty-five years' service, December 1986
Source: Private collection.

Figure 5.3. Celebration after the twenty-five years' service ceremony in a restaurant, Pula, December 1986
Source: Private collection.

challenges of constructing a ship of this size and how they rose to those challenges through a combination of individual and collective efforts. Beaming with joy and with obvious pride, they present the fruits of their labour.[50] Similarly, in the film *Berge Istra*, produced in the same year, the supertanker is portrayed as the common product of the entire collective – ultimately honoured by a visit from Josip Tito and his wife Jovanka, the vessel's "godmother," at the ship's naming ceremony.[51] At events like this, where new ships were launched, the universal recognition of the dockworkers is particularly tangible; these were public events that generally made the front pages of the local newspapers.[52] In his book, Bo Stråth remarks, "Ships are … the only industrial products besides aeroplanes which are the objects of christening. The industry is associated with pride and machismos; it is difficult to imagine textiles in such a role."[53] Thus, not only did the shipbuilders produce the ship, but the reverse was also true: the workers also became the objects of production.

Crucially, the workers' feeling of self-esteem was inextricably linked to the products of their labour. These were "not simply containers with an engine," but "highly advanced" products of exceptional quality that enjoyed worldwide recognition (B.Ba., one of our interviewees in Pula). This, in turn, created a connection between the workplace and the rest of the world.[54] There were few other products for which the official propaganda was so in tune with the emotions of the "ordinary people." In this context, government and company rhetoric used very similar arguments, presenting the shipyards as a trademark and thus as being among the few companies in the country to manufacture an internationally competitive product. Even the experts praised the quality of the ships.[55] *Glas Istre* never passed up an opportunity to highlight the technical sophistication of the ships and their commercial success abroad. From the workers' perspective, economic success and technical progress can be summarized as follows (quoted here from an interview conducted in Gdynia): "I was in my element. I built this massive thing, a ship like this, it was everything. I was a boy from the provinces, and when I arrived at the shipyard and began to build these ships.… It was like flying to the moon" (C.Z.). The company's self-representation also reflects this interplay between individual dedication and successful production.

The former workers of the shipyard in Pula are especially proud of their technological niche products, which include dredgers, "animal hotels" (transport vessels for livestock), specialized shallow-draught ships with collapsible roofs that can navigate rivers and travel under bridges all the way to the Caspian Sea, and a platform for the huge land

Figure 5.4. Welding work in Gdynia, 1973
Source: Courtesy of KARTA Center Foundation Archive, © Romuald Broniarek.

reclamation and real estate project "The Palms" in Dubai. These specialized constructions are repeatedly given as examples of the advanced engineering skills of the Uljanik shipyard and its workers. And the situation is similar on the Polish Baltic coast. One of our interlocutors even described the ships built in Gdynia as "floating Mercedes" (K.J.).

This sense of pride and personal achievement was articulated by men and women alike – in the aforementioned documentaries about Uljanik, female welders are also featured, an image rarely seen in Western shipyards. The female perspective adds another dimension: the ability to simultaneously function as a good worker and a good mother (M.D.). Yet, our female interlocutors made it clear that it was often difficult to reconcile family and working life, with overtime, for instance, frequently clashing with family obligations.[56] This well-known problem of the "double burden" that women workers had to bear under socialism, including Yugoslavia, has already been extensively researched. Although they readily described the difficulties, when asked directly, the women we interviewed still claimed they were generally very happy to have worked for the shipyard. This is likely, at least to some extent, to be related to the fact that the women we interviewed had mainly held white-collar jobs that were ranked higher than the ordinary workers in the social hierarchy (albeit not in the view of the shipyard workers themselves).[57]

The multifunctionality of the shipyards described meant the workers saw their lives as part of the collective effort to build a better future and consequently developed a specific concept of industrial modernity. At the same time, the shipyards also provided the infrastructural basis for the creation of a community and feeling of solidarity. One female worker from Uljanik told us that "the people showed great solidarity with one another" (J.L.). A typical example of this was how the workers helped each other with private renovation or construction work. Everyone we spoke to emphasized how easy it was to get support from co-workers. This process is similar to the one Muehlebach describes in her study of the workers' community in Sesto San Giovanni, Italy: "That labour and the productive process – the massing of tens of thousands of workers on shop floors, around machines, and in a densely crowded town – created the conditions not just for industrial output but for collectively held values as well."[58]

This sense of community also transcended the workforce. During the 1967 strike, the workers had thrown the management of Uljanik overboard,[59] but the new generation of directors, who took up their posts in the early 1970s, had a more respectful, "more compassionate" manner (J.L.). Unlike the old cadres, who (in J.L.'s words) were as "stern as a father," the younger managers appeared more laid back, spending their breaks with co-workers, chatting, and laughing with them – and all this helped cultivate a sense of being part of one big "family." Similar images of family-like relations in the workplace were also conjured up by the female textile workers Nina Vodopivec interviewed in her study on Slovenia. Vodopivec contends that the idealization of interpersonal relationships also served to restore, in narrative form, the real-world communities that had been shattered by transformation.[60]

The family metaphor repeatedly crops up in the interviews we conducted in Gdynia as well. For instance, this familiarity found expression in the strike that took place in summer 1980, where, according to one interviewee "we stood as one big family" (C.Z.). Indeed, the majority of historical analyses of labour struggles emphasize the strong personal and familial dimension of the opposition movement at that time. The fact that many of the workers were neighbours in residential districts owned or built by the shipyard, that their children played in the same playground and sat next to one another in school served to reinforce the feeling of communality. Another important place of "togetherness" was the bus or train on which thousands made their daily commute. This included, for instance, the local train that connected Gdańsk, Sopot, and Gdynia and came to symbolize the lives and protests of the dockworkers.[61]

Rifts and Compromises: New Social Values and Economic Practices

During the 1990s, the once inextricable bonds between the workers' lives and the shipyard gradually began to come undone, causing important channels for disseminating information to dry up. In 1991, the company newspaper *Uljanik*, which had appeared monthly and, in its heyday, comprised three hundred densely printed and illustrated pages a year, ceased publication. This was a result not only of a serious lack of funds but also of the imminent end of self-management. Now that the collective no longer managed the company itself, there was no need to keep the workers so well informed about the relevant processes and goings on. All that remained was the thin, small-format information bulletin *Mali informator*, which no longer gave the staff or the shipyard's multiple cultural and sports organizations an opportunity to air their concerns. With the end of socialism, the company increasingly became something of a "black box." The public now knew very little about what went on in the production halls, and any information they did receive was filtered through the PR department. Public relations replaced genuine information.

In other areas too, the workers experienced a loss of control, almost a kind of dispossession. One of our interlocutors described how expectations had changed: "[Today] the workers are totally exploited. It wasn't like that before" (N.T.). The pace of work was not sustainable and over the years exhausted the workers, resulting in dissatisfaction. In the words of one Croatian trade union official, work during the self-management era was much more laid back – in those days "there were more excursions and less training" (B.Bu.).

In Poland the workers resisted such attempts to increase the pace of production, a move that would have restricted their spaces of autonomy. One of the former so-called "rationalizers" (E.P.) told us about a proposal made in the 1990s aimed at optimizing lunchbreaks for the hundreds workers at the dry dock, which was located more than a kilometre from the canteen (Pula had a more practical set-up as the final assembly department on Olive Island had its own canteen). The walk to the canteen and back took half an hour, lunch took another half an hour, meaning that the workers' lunch break – which the company paid them for – took at least an hour, often more. To our interlocutor, the solution seemed obvious: a big, heated tent would be constructed right next to the dock and lunch would be served from there. All calculations suggested that this would be a very profitable investment. The solution would shorten workers' breaks but also make life easier for them, which, in turn, would have meant increased productivity. Unfortunately, however, according to our interviewee, this pragmatic idea was rejected by the workers and ultimately not implemented (E.P.). It is not hard to figure out that the source of this opposition was a fear of stricter control and increased work discipline.

The "fantastic camaraderie" of the past gradually disappeared, partly due to the growing disparities between the workers. One key reason for this was the increasing flexibilization of employment relations, for instance through the use of agency workers and subcontractors. In 1999, for example, there were 1,865 employees and 850 subcontractors or temporary agency workers (*kooperanti*) building ships in Uljanik.[62] The documentary *Godine hrđe* (The years of rust) produced in 1999 by Pula-born director Andrej Korovljev, who was twenty-eight at the time, used striking images to show that, for the *kooperanti*, working in the shipyard was no more than a very hard way of making a living, and produced neither meaning nor solidarity – quite the opposite, in fact. The analysis of the documentary by ethnologist Andrea Matošević concludes that the film portrays the workers employed via subcontractors as entirely alienated from the products of their labour, no longer the subject in the production process. There is no longer any trace of pride, with resentment prevailing instead.[63] These agency workers carried out the most dangerous jobs (such as applying corrosion protection to the ship's hull), they earned less than regular employees, and were not entitled to company social benefits, as Matošević's fieldwork in the shipyard showed. They were expected to produce only quantity, not quality, which explains why it was impossible for them to positively identify with the shipyard. In the eyes of the regular employees, the agency staff symbolized the threat that they too would be "flexibilized."

After the turn of the millennium, Gdynia saw a wave of rationalization that sparked protest among the workers. Interestingly, this was hardly mentioned in our interviews, something that can, to a certain extent, be explained by sample bias: almost all of our interlocutors worked directly in production. From their perspective, the financial and structural upheavals played out somewhere above their heads or in other departments and so had little impact on their day-to-day activities. Their focus was entirely on solving technical problems and building ships to be sold on the world market. Some of our interviewees did raise the issue of internal restructuring, but this was mainly depicted as formal reorganization that did not influence the "real" work to any great extent (E.S.). Even the workers who were transferred to the shipyard's subcontractors described this continuity. For instance, at the time of the interview, one of our interlocutors was working for a company that made scaffolding for the regional shipbuilding industry – and also for numerous cultural events in Gdynia and the surrounding area. Essentially, he was still doing the same job in the same place, just in a completely different context and with much less potential for the production of meaning.

In Uljanik, when it comes to the experience of day-to-day work, narratives of discontinuity are more prevalent today. According to our interviewees, the changed relations between workers was reflected in the fact that in the past, it would have been inconceivable for someone to retire without an informal celebration, kind words, and gifts. Now (meaning the last few years before the shipyard was finally closed), nothing of the kind happens anymore, and it is not unheard of for people to leave the jobs they have done their whole lives without anyone even noticing (N.T.). In the view of the workers we spoke to, the solidarity ascribed to the past had now disappeared completely (L.Ž.). The one thing that still seemed to drive people – both politically and professionally – was better pay, a former trade union official reported (B.Bu.). Moreover, the once collective struggle for higher wages had become more highly individualized. The wage spread by department and qualification aroused envy and resentment, which was further reinforced by the meritocratic rhetoric of the neoliberal 1990s.

In the end, as all the managers and workers we interviewed consistently confirmed, relations between the workforce and the management changed dramatically. "In the past you could go and talk to the director any time, and you knew he would help you." Nowadays, if you ask "why something is done a certain way," "the answer is always 'That's just the way it is!' and see you later" (Z.M.). These days, it would never enter anyone's head to knock on the manager's door. Z.M. then went on to say that the management of the company now had a very different character, having become far "more rigid." What was once internalized discipline became externalized. The arrival of the new post-socialist managers meant the establishment of a new leadership culture, which, in the workers' eyes was characterized by disdain for the workforce. This led to even greater disenchantment with and alienation from the shipyard and its work. A very real consequence of this change repeatedly highlighted by our interlocutors was that the communication between the heads of departments and the workers was no longer "normal"; it was no longer possible to have "rational discussions" about work issues and the only rationale behind relations was "survival of the fittest" (B.Bu.). Even discussions about work conditions were tersely dismissed by the new managers: "If you don't want to work under these conditions, then you know where the door is." If you saw your boss from afar, you did your utmost "to adeptly stay out of his way, to avoid coming into contact with him," at least this was the impression shared by one of our interlocutors in Pula (J.P.).

In the past, it was not only hierarchies and formal rules that shaped the workers' relationship with their supervisors, but, more importantly, respect. One of the workers we interviewed in Pula observed, "We had great respect for our boss because he was really someone who had earned his position" (J.L., mainly in reference to the 1960s and 1970s). This was a widely held view in Pula and stemmed from the fact that the general director, like all other directors, was appointed through an internal selection procedure, which meant he tended to have cut his professional teeth in the shipyard itself. In other words, "We trained our directors ourselves, we shaped them over the course of the production process" (R.C.). No-one was ever appointed from outside; the public respect for the director and the recognition he was given for his expertise and skills were considered crucial for the running of the company.[64] The long-standing director Karlo Radolović is still a local hero, celebrated by many of the workers we interviewed (and the local press) for having skilfully rescued the shipyard, or at least having prevented it from being "sold off." Other directors are described as similarly "savvy" (L.Ž.), although their overall standing already began to crumble in the early 1990s.[65]

In Gdynia, the local hero from the 1980s was Zbigniew Maciejewski. A week after the imposition of martial law in December 1981, Maciejewski, who was formerly deputy technical director and then deputy director for production, took over the management of the shipyard. "This guy was first a worker, then a foreman, then a master craftsman" (E.P.). First and foremost, Maciejewski was a shipbuilding engineer who had studied in Gdańsk and knew production like the back of his hand. The older shipyard workers repeatedly portrayed him as the antithesis to the post-socialist manager: "This director arrived asking where the ship's bow was, and the stern. Good god, this guy was clueless. This Maciejewski, they need to bring him back to life, he would be celebrated, he would figure something out" (E.P.). The Polish verb *wykombinować* (to figure out) used by E.P. refers in this case to the leeway the director had under socialism and the informal practices underpinning his influence.

In Pula, unlike in Gdynia, no high-level managers were ever appointed from outside – not even in the post-socialist period. Instead Uljanik tended to rely on directors with a pedigree. All our interviewees were convinced that it was this very continuity of managers, experiences, and skills – Radolović being a prime example – that was the saving grace of the shipyard during the period of transformation. Accordingly, the abrupt break in this continuity, associated with the end of the shipyard's practice of training its own skilled workers and managers, was seen as the decisive factor in its demise (S.B.).[66] These assessments once again provide a glimpse of the ideal of the holistic company, an ideal that does not fit with the neoliberal idea of business economics. Interestingly, the director who led Uljanik into bankruptcy, Gianni Rossanda, was never afforded this degree of respect, never recognized as a true member of the Uljanik family, despite the fact that he came from Pula and he, too, had many years of service under his belt, even working for the shipyard as a student. Rossanda was, however, an economist, heading Uljanik's finance department for a time, which is why he was often symbolically associated with the unpopular financialization of the company.[67]

This change in attitude towards the company, the growing scepticism regarding its management, and the loss of identification with the ships was also related to the fact that the shipyard was increasingly disappearing from the public eye (following privatization) and the public had started to lose interest in it. This public recognition, once such a crucial basis for the pride felt by the workers, disappeared. The population of Pula only peripherally acknowledged launches of new ships, sometimes even taking no notice of them at all, as one of the electricians we interviewed told us with sadness (Z.M.). A similar, perhaps even stronger sense of loss is evident in Gdynia, where the infrastructure of the former Paris Commune Shipyard or Stocznia Gdynia is used to this day. Many of the shipbuilding companies operating on the grounds of the former shipyard today, having achieved success in niches of the European and global markets, conduct their business in silence, practically invisible. They are closed to the outside world, the visual spectacle of production is concealed, not only to the tourists visiting the city but also to the majority of Gdynia's population. Today, the shipbuilding industry has been symbolically marginalized on the urban landscape – and in the "landscape of consciousness" all the more so. In Pula, in contrast, the motionless cranes and the torsos of unfinished ships on the outskirts of the old town serve as a daily reminder of the shipyard's demise. For years, the city skyline was marred by the huge cargo ship *Santiago II*, anchored, rusting away, not 100 metres from the Temple of Augustus. Although the shipyard had almost completed this massive vessel, the customer went bankrupt and no one was interested in taking the ship off their hands. And to this day the carcase of a huge livestock transporter adorns the dock on Olive Island in the middle of the Bay of Pula.

New Horizons on the Old Industrial Landscapes

So, are things really this bleak? Is there truly nothing left of this ethos and myth of shipbuilding, which feels so long ago, yet so very present at the same time? This is where our two case studies differ. The collapse of Stocznia Gdynia in 2009 can be seen as a symbolic moment that heralded the end of a type of state-organized industrial lifeworld. This was a socialist version of modernity, prolonged in post-socialism and constructed around labour, industrial labour in particular. The relative success of Poland's (private) shipbuilding industry after this date is undisputed, but this is already an example of a whole other form of "modernization."

Crist employees, along with the numerous agency and subcontractor workers, can be proud of their technologically advanced and highly competitive products. However, the company is not interested in adding any value in terms of the social and individual or personal meaning of those products. This is not something that is necessary from an economic perspective; nor does anyone expect it – not even the employees themselves. And our interviewees, many of whom still worked for these companies and were perfectly aware of the radical change, had a similar perception of the situation. That is not to say that the employees are entirely unhappy with their current working conditions or do not identify with the products they are manufacturing. Their private satisfaction, however, has no broader meaning for the community, the city, or the country as a whole, and is no longer the basis for far-reaching visions of a good society. These meanings have become mere nostalgic memories that are barely present in the institutionalized public consciousness.

Today, thanks to its size and the importance of tourism, Gdynia is a city with a lively cultural life, with several newly established or revamped museums, the most popular of which are the Museum of Emigration, the City Museum, and the Naval Museum. Not one of these museums devotes very much attention to the work of the shipyards that once characterized the city (albeit never completely) for decades.[68] Gdynia tries hard to present itself as a modern city – the modernist architecture of the interwar years lends itself better to this undertaking – disregarding its socialist, industry-oriented modernization "episode." This is an observation, not a complaint. At the end of the day, the way in which the city presents itself is geared towards the wishes of its consumers.

In Istria, the shipyard continued to produce symbolic "added value" for almost a decade longer than in Gdynia. Uljanik is located in the centre of Pula, its products were almost always visible to the residents and tourists, and up until early 2018 left many with the general impression that the shipyard had successfully adapted to the new global economic cycles. The association of the shipyard with modernity – this time the post-socialist version – appeared to be intact, embedded in the contemporary rhetoric of global competitiveness. As recently as 2017, the shipyard's employees still found many reasons to proudly express their sense of self-esteem. For example, one of our interlocutors, who worked in the marketing department selling electric machinery and equipment, told us, "We are actually the only [Croatian] shipyard in the north [of the Eastern Adriatic] left today…. We build ships with more added value, there is space, we have our niche" (S.B.). He then went on to list the competitive products currently being manufactured in Uljanik: platforms for natural gas extraction, dredgers for the construction of the world-famous Palm Islands in Dubai, and so on.

Today, six years since the shipyard went bankrupt, an attempt to redefine its spectacular remnants is underway. Some would prefer to see the factory buildings torn down to make space for hotels and a yacht marina, and to liberate the city from the memory of socialist-era heavy industry once and for all. Others are contemplating how to conserve the remains of the shipyard, perhaps "valorizing" the "industrial heritage," preserving it as part of the collective memory, even making it something of a tourist attraction. Yet, in neither of these scenarios do the memories and emotions of the erstwhile shipyard workers play a role. They are a mere object of negation or museumization. How, therefore, can we conceptualize – beyond rhetoric and top-down strategies of representation – the day-to-day reappropriations and forward-looking life strategies we observed in the city?

In seeking an answer to this question, we came across the excellent work of Jeremy Morris. This social anthropologist, currently a professor at Aarhus University in Denmark, published a study on post-socialism in a central Russian industrial town where the main employer had gone bankrupt. In his study, Morris coins the phrase "habitability" to emphasize that those people who were once members of the working class should not be seen as mere passive victims of neoliberal developments – and places like this Russian monotown should not be stigmatized as decaying, half-dead relics of the past struggling for sheer survival. In fact, the one-time heroes of the industrial workforce are actively doing something, not only in a vague attempt to find their bearings under what are granted very difficult material circumstances, but to actually develop strategies so that one day they, or at least their children, will have better lives. And here a lot of their old habits and skills stand them in good stead, including practices like reciprocity among co-workers and neighbours, the ability to repair things themselves or use informal channels to get things "done," as well as a certain imperviousness to adverse conditions.[69]

In Pula, too, the former workers demonstrated – just as they had done when they worked for the shipyard – their capacity to pursue their interests as best they could in the face of the many trials and tribulations. They showed that they are able to change and adapt in order to achieve this, and to make the places where they lived "habitable" – that is, comfortable enough so that they could enjoy a "good" or "normal" life. The residential blocks of former company-owned apartments are in good shape (unlike those in many de-industrialized towns in the West). The front gardens are well kempt – after all the apartments now belong to the former workers. In summer, the flats are rented out to tourists and the residents' days are spent frequenting the nearby cafés to while away time. Listening to the workers in Pula – who show no trace of revolutionary, utopian, or other ambitious ideologies whatsoever – they are just trying to come to terms with their new circumstances and live a decent life under the circumstances. "OK, so the pension is what it is. It's no higher than the Croatian average. OK, well … the pension is definitely modest, but I'm not complaining" (Č.Č.). This is certainly not an expression of pure fatalism, but rather a conscious attempt to get to grips with the present and, step by step, find ways of making life more habitable instead of languishing in nostalgic memories of the good old days, which would serve only to make you feel worse.

Uljanik in its post-socialist form and the society that grew up around it are therefore not a lesson on decline and disintegration. No more than the de-industrialized city of Pula can plausibly be seen as a ruin – much like Gdynia, it is still a perfectly pleasant place to live. In this sense, the protracted process of downsizing, privatization, and the fragmentation of the two shipyards, which dragged on for years, meant that the gaps they left after their bankruptcy – whether socio-economic, symbolic, or in the life stories of the workers – were far less significant than the tried and tested narratives used by their directors and trade union representatives to legitimize requests for state aid would have us believe. In other words, when the shipyards stopped conducting their core business, they sometimes left no gap at all, as the social and physical infrastructure (which had previously been seen as an obstacle to the "core business") survived the bankruptcies, at least to some degree.

In Gdynia, the old infrastructure continues to be heavily used although no-one makes a big deal of it (figure 5.5). To some extent, this continuity is similar to the structural change in the Western European shipyards and steel-producing regions. Although these industries no longer play the role they once did, economically, in terms of employment, and especially in public consciousness, they do continue to exist in certain niches. This certainly helped mitigate the effects of structural change, despite the significant acceleration of the latter after 1989, particularly after EU expansion. This was, however, less the result of active labour market policy or massive state investment, as was the case in the EU member states in the 1980s and 1990s, but more owing to the dynamics of the private sector. The transition was also cushioned by the rapid growth of the tertiary sector (predominantly driven by the catch-up demand in all manner of services), the EU subsidies that Poland and Croatia received in abundance, and last but not least, the development of the tourist industry, which was of course not possible in the less picturesque locations. The latter were hit by the full force of deindustrialization, facing consequences of an entirely different quality altogether when it came to the prosperity, self-confidence, and development opportunities of the various municipalities. This is something that can be clearly seen in many small and medium-sized towns in the north of England or the American Rust Belt. However, the media images of post-industrial bleakness can sometimes be misleading, with people continuing to try to make their lives more bearable even atop the material ruins of industrial modernity.[70] Another factor to bear in mind is that people in Eastern Europe may always have had fewer illusions about the stability of their modest prosperity than workers in the West, who had internalized the capitalist promise of consumerism.

The aforementioned habitability depends on social ties and relationships built on the remnants of the past – both material, such as the inherited infrastructure, and intangible remnants like work cultures and identities.[71] The legacy of the recent past is just as apparent at the individual and family level: people live in apartments constructed by the shipyard or houses they built with subsidized loans the shipyard provided. If you visit their homes, you will see framed ship construction

Figure 5.5. Crist shipyard, 2020
Source: Photograph by Piotr Filipkowski.

diplomas belonging to their fathers and grandfathers on the walls, and proudly displayed in the cabinets are the awards they received for being exemplary workers. These remnants are more than just relics of the past as they continue to shape the daily lives of these people, both symbolically and palpably.

Beyond the workers' homes, on the collective level, too, people are seeking to make their lives habitable. Even after it went bankrupt, the Uljanik shipyard in Pula continued to occupy an important and predominantly positive place in the collective landscapes of consciousness. Its presence transcends nostalgic memories: the lion's share of the social and cultural infrastructure that Uljanik developed in the socialist period still exists and indeed is still in use, even though it is no longer owned or managed by the now idle shipyard. The Uljanik stadium and sports ground in Stoja, the (renamed) football club, the Uljanik nightclub, the eponymous club for pensioners, the brightly illuminated shipyard cranes, the yacht club, the fishing club – all these facilities have kept the name Uljanik alive, and not only in the public memory. The opening of the elegant rustic-style "Shipyard Pub" in 2016 leads us to assume that even commercial enterprises know how to use the sentimental association with the shipbuilding tradition to their own advantage. The shipyard, referred to in local dialect by the deferential term *škver*, from the Venetian word for shipyard (*squero*), which, in turn, stems from the Greek *eskhareîon* (hearth), has become an integral part of the city's official narrative and is used as a positive distinguishing feature. In physical terms, the shipyard was and still is inextricably bound to the city. Even the municipal authorities have discovered its use as a city marketing tool. In 2014, Dean Skira, a renowned local designer and owner of a lighting company in Pula, secured backing from the city, the Ministry of Tourism, the biggest local tourist company, and the Uljanik shipyard itself for his "Lightning Giants" project, which saw the huge shipyard cranes illuminated with coloured spotlights at night.[72] This became an overnight tourist attraction and the light show can still be seen today.

Former Uljanik workers, their children and families, and even those citizens of Pula who were not directly employed in shipbuilding still live in a city that is strongly defined by the erstwhile *grandezza* of the shipyard. Many have developed adaptation strategies and have managed to tame the far-reaching social and economic processes that pulled the ground from under them. Thus, below the surface of the omnipresent narrative of decline, new meanings are produced, albeit not yet publicly articulated. It seems as though people are slowly beginning to accept that Uljanik is no more (although, as of 2022, there was still a stubborn core of workers who were attempting to revive production; a project the city's population regarded with great empathy, if little optimism, and which has so far come to very little by way of visible results). Whatever eventually emerges from the old Uljanik plant, it will certainly be something new. The city of Gdynia, in contrast, is already endeavouring to symbolically dismantle its "giants," so as not to be burdened with this ambivalent baggage of the past as it moves forward into the future.

Abandoning the big shipyard was not easy – neither mentally, nor socially. It was not something that happened overnight but rather as part of a long process. The long-standing shipyards in Pula and Gdynia succeeded in preserving their industrial lifeworlds way beyond the collapse of state socialism – something many (post-)socialist companies did not manage. For a while it even seemed as though the shipyards would be successful in their attempts to adapt to the new rules of the market economy and international competition. The majority of Western European shipyards had already been shut down decades earlier, as the lyrics of Sting's song bear witness to. Yet, even before the two Eastern European shipyards at the centre of this study went bankrupt, the personal and social importance of working in shipbuilding and other industries was all but gone. It was now a private affair, no longer a collective value, and had stopped serving as a point of reference for public creation of meaning.

In the past, as part of the socialist modernization project, the two shipyards had created a surplus of meaning, in a sense as compensation for their low financial return. For the workers, this meaning and the associated social recognition represented a kind of reward for their hard work and, compared to the West, rather modest wages. Had their work not been celebrated and symbolically exaggerated in this way, the modernization project would presumably have failed much earlier. In the end, this arrangement went to rack and ruin because of the massively inefficient production, the resulting decline in global competitiveness, and the loss of political legitimation. However, this experience remains part of many people's memories; for many, in fact, their only experience of the world of work. To quote Sting again, "The only life we've ever known is in the shipyard" (from the track "Shipyard" on the aforementioned album *The Last Ship*).

Ultimately, the rocky marriage between the rationales of old and new modernization agendas proved to be unsustainable and fell apart. While for Pula, the divorce settlement envisaged its transformation into a tourist monocity, Gdynia came up with a more pragmatic solution. With its still significant – albeit "invisible" – shipbuilding industry, the growing container and passenger port, thriving cultural life, and high ranking in polls on "life satisfaction,"[73] the city is in a position to market itself as a role model for successful transition from industrial to postindustrial modernity. Interestingly, however, the latter nurtures a certain nostalgia for the former (albeit mainly among the older generation) and continues to use much of its "socialist" infrastructure and expertise, if only on the quiet.

6 Keel Up? The Future of Shipbuilding in the EU

On 3 April 2020, the *Hua Yang Long*, a 52,000 DWT semi-submersible heavy cargo vessel measuring 228 metres in length and 43 metres in width, docked in the Bay of Pula. The ship, the property of the Guangzhou Salvage Bureau, a company directly under the Ministry of Transport of the People's Republic of China, was tasked with the unusual job of picking up a special shipment at the Uljanik shipyard – a self-propelled cutter suction dredger, a vessel whose construction was only 80 per cent complete when the shipyard went bankrupt. The *Willem van Rubroeck* dredger came in at an impressive 151 metres long and boasted a cutter capacity of 8,500 kW and a dredging depth of up to 45 metres, making it the most powerful vessel of its kind in the world. Indeed, the vessel was once the jewel in the crown of Uljanik's order book. With a contract value of some 800 million kuna (around 108 million euros), the dredger would also have been the most expensive ship ever to be built by a Croatian shipyard.[1] According to the order, which had been placed by the Luxemburg-based maritime construction company Jan de Nul Group, completion of the vessel was scheduled for autumn 2018.[2] Instead, however, the *Hua Yang Long* took what was now the *Willem van Rubroeck* on board to transport it from Pula to the Remontowa shipyard in Gdańsk – a shipyard specializing in repairs. Interestingly, this shipyard is one of the seldom mentioned success stories associated with the restructuring of Poland's shipbuilding industry.[3]

In many respects, this episode sums up the very story that our book is endeavouring to recount (although the proverbial irony of history did not favour the authors and allow the successor company operating Gdynia Shipyard to complete construction on the *Willem van Rubroek*). Uljanik, unable to survive under the conditions created by the rigorous EU competition laws, ultimately went bankrupt in 2018. The management failed to see the writing on the wall from Brussels, instead continuing to rely on their typical practice of muddling through. Still confident that it would receive state guarantees, the shipyard took on multiple orders, despite knowing from the outset that it would not be able to deliver on schedule. Clearly, neither the shipyard management nor the Croatian government had grasped just how much Croatia's accession to the European Union had transformed state-business relations. In fact, they did not really learn this lesson until additional state aid for the loss-making shipyard was vetoed by Brussels. And yet, Croatian economic policymakers could have learned from the precedent set by Gdynia Shipyard nine years earlier (2009). At that time, as already explained, the European Commission had instructed the Polish government to recover the subsidies they had granted contrary to EU state aid law. This intervention was to be the end of the shipyard on the Baltic coast.

But there is another lesson to be learned from Gdynia, with clear signs of a shipbuilding revival a decade later, albeit on a smaller scale and with narrower product portfolios. The situation in Gdynia and in Gdańsk suggests that the rules set by the European single market have not wiped out shipbuilding in Europe altogether. From the ruins of the giant socialist shipyards, new specialized companies that are continuing the production of key ship components are emerging. These successor companies rely on the local maritime industry ecosystems inherited from the socialist era, including port infrastructure, skilled labour, technical colleges, the good reputation of the shipyards, and the remaining docks and cranes.

Today, however, Polish shipbuilding is quite different from the holistic assembly process emblematic of socialism. Integration into global capitalism is evident in many areas, in the dubious practices of employing North Korean and Ukrainian workers, the plethora of subcontractors and temporary work agencies, and the almost Babylonian diversity of languages heard on the slipways. These shifts reflect the characteristics of the revival of industry in Eastern Europe, a region that has managed to secure an important place in transnational supply chains and production networks thanks to a combination of cheap yet skilled labour, good infrastructure, proximity to Europe's centres of wealth, large-scale foreign investment, as well as integration into the European single market and access to EU structural aid. In recent years, in fact, the share of manufacturing in total employment has even increased in Eastern Europe, challenging the prevailing narrative of complete deindustrialization in post-socialist Europe. In Poland and Croatia, more people are now employed in industry (in terms of their share in total employment) than in Germany and significantly more than in the birthplace of the Industrial Revolution, Great Britain.[4]

China and the Situatedness of the Transformation

The episode with the Chinese freighter illustrates another lasting change that has shaped shipbuilding – and the world at large – in the last three decades and will determine the future: the rise of the PRC, the largest shipbuilding nation in the world and at once an industrial and financial giant. China is showing just how determined it is to take on a leading position in the construction of merchant ships of all kinds. Between 2003 and 2015 alone, the OECD reported fourteen Chinese government development plans in which the shipbuilding industry was involved or received special mention.[5] From very humble beginnings, Chinese shipbuilding emerged as a major player in global production and distribution networks. By the second decade of the twenty-first century, China was already the world's largest shipbuilding nation, accounting for around 35 per cent of orders (in tonnage), followed by South Korea (also more than 30 per cent), and Japan (slightly less than 20 per cent), while the EU as a whole came in at as little as around 5 per cent.[6] There is nothing to suggest that the dominant position held by the East Asian shipbuilding companies is likely to change in the near future. The fact that the *Hua Yang Long* travelled many thousands of nautical miles from China to transport a European-built vessel from one EU shipyard to the next can be taken as proof of China's successful entry to the market for highly specialized, technologically complex vessels.

Is there a better symbol for the forces that Europe's comparatively small shipyards are up against? Besides unparalleled economies of scale and extensive technological links with other industries, Chinese shipyards also enjoy the support of a government that has never abandoned "soft budget constraints" and an active industrial policy, especially since many Chinese shipyards are state owned.[7] Moreover, thanks to the bloated Chinese steel industry, Chinese producers benefit from lower steel prices, an important cost factor in shipbuilding.[8] Another (unfair) advantage that Chinese companies have is the poor state of workers' rights in the PRC, as the repressive state does not allow free trade unions to operate. However, the Chinese government does not distribute its support indiscriminately or haphazardly, but assists those shipbuilders who are competitive, while insolvent shipbuilding firms do not benefit from its generosity. The Chinese government is also not afraid to use foreign policy levers to boost the commercial success of its national companies. On the (south)eastern border of the EU, China is increasingly materializing as an investor or lender with deep pockets – interestingly the country seems particularly keen to acquire critical infrastructures such as ports and railway lines that are to be prepared for the much-vaunted "Belt and Road Initiative." It seems as if China might even succeed in doing what the Eastern European state socialist countries were denied the opportunity to do – namely, creating an authoritarian developmental state that outstrips the West in the technology race. Kornai's observations on the inherent inefficiency of the state entrepreneur need to be re-evaluated for the case of China.

The growing importance of China for the fate of European shipbuilding illustrates one of the key findings of our research, that is, the close interconnections between changes at the global and local level, with the nation state as well as mid-level institutions such as municipalities and industry associations mediating between these levels and their dynamics. These interconnections are one reason why the transformation in Eastern Europe took a very different shape than the structural change from heavy industry to services and information technology in Western Europe. What is clear is that the communists' unshakeable ideological penchant for smokestacks and blast furnaces was the main reason why Eastern European economies could not bring themselves to leave coal and steel behind as early as the 1970s and 1980s, although this might have been the most logical move given the impact of the oil price shock.[9] Only when the socialist economies finally ran out of steam and the governments could no longer conceal their growing losses – or compensate for them by borrowing from the West – did they or their successors implement belated economic reforms, and when these reforms did come, they were all the more radical.

As ambiguous, often contradictory, meandering, and controversial as the reforms in the course of the transition to a market economy were[10] – when it comes to the specific trajectory of the transition as well as the outcome, the timing was to prove as decisive as the geographical location. First, it was not only the Eastern European economies that suddenly entered the world market with full force at the beginning of the 1990s, integrating their workforce into the global division of labour, but also China with its even greater resources and rapidly growing industrial capacities. The international division of labour put the new EU member states in a precarious intermediate position. Although they were able to compete with the economies of the old EU countries, especially in terms of labour costs, in relation to China and East Asia as a whole, this held true only to a limited extent. The countries of the West, in contrast, did not face the same twofold challenge during their structural transformation.

Second, the ever-greater integration of Europe's common market from the mid-1980s onwards fundamentally changed the economic policy paradigm – the rules of the game were different now, also and especially for states that joined later. While, in response to the two oil price and economic crises of 1974 and 1979, Western European governments were still able to provide quite generous aid to their domestic shipbuilding industry, after 1981 increasingly strict state aid laws were introduced.[11] The headway made with European integration, which was happening parallel to the process of transformation in post-socialist countries, resulted in a huge shift in regulatory competences to Brussels – without the political actors and general public in the individual member states or candidate countries being fully aware of this. These factors meant that the transition economies had to position themselves in a world that looked very different from the one that had existed at the beginning of deindustrialization in Western Europe in the early 1970s. The Western European producers that were pushing their way into low-wage countries in the 1990s and the banks who were looking for investment opportunities quickly discovered the potential of Eastern Europe, even bringing about a veritable re-industrialization in some locations – albeit at the cost of integration at the lower end of the international value chain, where production could often be relocated to even cheaper countries. Moreover, the shock therapy involved enormous social costs, especially in the early 1990s. The high degree of dependence on foreign capital inflows made the economies in the region vulnerable – in sectors such as shipbuilding where foreign capital failed to materialize, re-industrialization proved difficult.[12]

This leads us to the third major difference between late or post-socialist transformation and the capitalist pattern of structural change seen in Western Europe. The latter was accompanied by considerable state mitigation. And while this did not go far towards overcoming the crisis in heavy industry after the second oil price shock, what it did do was help companies and some of their workers to continue production in specialized niches and train the workforce needed for these areas (the two cradles of neoliberalism, the United States and Britain, had less of a hand in the direction of structural change with corresponding social dislocation).[13] Moreover, until the mid-1980s, in Western European heavy industry, the companies themselves still had the means to invest and thus modernize. Trade unions and works councils often made constructive contributions here, although their primary interest lay in securing redundancy plans that were as generous as possible.

Post-socialist governments, in contrast, relied on procrastination tactics and subsidies in order not to lose too many votes and legitimacy. With competitive pressure on the world market from Chinese companies becoming increasingly noticeable from the 1990s on, Eastern European states tried to support their shipyards by injecting capital and cancelling debts. But compared to China (and South Korea), they simply did not have sufficient funds to really help the companies, let alone for a technology catch-up. For the majority of the locations what this ultimately meant was a structural "clear-out," that is closure of a number of firms, although this did open up leeway for new private companies such as Crist in Gdynia or the repair yard in Gdańsk. The workers in Poland and Croatia who had lost their jobs or who were about to be laid off received severance pay, but very little in the way of retraining – their main financial resource was the housing that had been privatized in the 1990s. The younger generation switched to other industries or just moved to the West. In this respect, the workforce for a large-scale revival of the industry will also be lacking, should this one day be desired by the powers that be. This is something else the region has in common with the United States and Great Britain.

Thus, the comparison made here does not substantiate the well-known formula of the East-West dichotomy, which has been used time and again in all kinds of studies, most recently at the political level by Ivan Krastev and Stephen Holmes, who claim that the post-socialist countries first tried to imitate the Western model and, when that failed, turned away from the West.[14] The differences between Poland and Croatia are simply too great for the dichotomy to be of value, as are those between the individual countries of the former West.

The Hidden Liveability of Transformation

The workers we interviewed for this book are keenly aware of a fundamental contrast between "before" and "after," although they would have difficulty pinpointing exactly when one phase ends and the other begins. Only a few of them explicitly referred to the end of communist rule. This may well indicate that the socialism people knew from their everyday lives, on both sides of the factory gates, was defined and experienced using very different categories than those applied by the political system.

In the interviews, "privatization" was often described as cutting far more deeply than the political upheaval, although its meaning is as vague as its chronology. The use of the term by our interviewees does not necessarily denote a transformation of ownership. People referred to "privatization," for example, when talking about the shipyards still (or once again) being in the hands of the state in the 1990s or even later. Due to the loss of co-determination rights, as was the case with the dissolution of workers' self-management in the Uljanik shipyard, even nationalization can be experienced as privatization – as a form of expropriation.[15] What continued to affect the interviewees, even many years after the event, was the gradual and occasionally abrupt change in their work lives, daily routines, and social relations. This was the result of a new style of management that made old activities obsolete, implemented new rules, and increased work pressure. The sudden enforcement of safety regulations and the all-encompassing *papirologija*, as one of our interlocutors in Pula called the shift from autonomous to rule-based work practices with its extensive paperwork requirements (B.Ba.), mark the widespread feeling of a loss of autonomy.

Control over one's own working hours was the flip side of socialist inefficiency at a time when production depended even more on workers' problem-solving skills and their ability to improvise. In the course of the 1990s, the workers' practical knowledge became less important than abstract knowledge based on rules and theories – this was another aspect that our interviewees saw as falling under "privatization." The place of agency shifted to a space beyond the factory floor and the knowledge it embodied. Here, our interviewees' answers and our oral sources were much more multifaceted and, even with their lack of chronological definition, more accurate than the written sources that reduced privatization to a mere legal process and change of ownership. Oral history also makes the paradigm of individualization seem questionable, because individually, after the end of "actually existing socialism," what the workers – and women workers in particular – experienced was that they had less agency in – dare we say – "actually existing capitalism."

The workers described with regret the decline of the shipyards, which cut back many of their extensive welfare functions in the 1990s and shed many thousands of jobs. They accepted this without complaining about the unbridled misery, without developing or even permitting a narrative of social catastrophe – neither Gdynia nor Pula have turned into social hot spots. Yet a whole host of feelings and unsettling experiences gave them cause for complaint: loss of recognition, wounded pride, and alienation from work, whether they were still working in shipbuilding or already in other industries. Many people expected the shipyard to remain an idealized holistic "family," just as it had been under socialism. But most of the interviewees more or less accepted the new reality, albeit accompanied by a feeling of resignation: from now on, the shipyard was "just" a job that paid a comparatively good wage (which, in the early 1990s, was a privilege those in other companies and sectors did not enjoy). In the workers' recollections there are numerous references to the significant dismantling of the shipyard's community functions; the imaginary "we" of the Fordist socialist factory community was dissolved. In the factory halls, privatization presented itself in such a way that the private purpose overrode the purposes of the community – the symbolic congruence between shipyard work and way of life was no more.

One constant in the constructions of subjectivity of the shipyard workers, both former and active, is the feeling of identification with the fruits of their labour, a constant that is manifest in the ships that towered over the silhouette of the city and everyday life. The overdetermination of the physical objects explains why the emotional void left by the insolvent shipyards is so deep for the former workers and their local communities. The liquidation of the shipyards meant more than a loss of jobs or wages; more importantly, it represents the separation of the self from a tangible outcome of collective effort and from a wider world that people once felt connected to through the product of their labour. After all, even in the most difficult times of the late 1980s / early 1990s, the workers continued to build ships. The fact that, with the economy as a whole having overcome the transformation crisis of the early 1990s and the much-celebrated accession to the EU successfully achieved, this was no longer the case, is perceived by many as a downright personal insult. In addition to the unfulfilled promise of progress, this is another reason why, unlike structural change in the West, the transformation in the East resulted in such a crisis of political legitimacy.

We interpret the fact that the workers express the idea of community and collectivism so clearly in their recollections not only as a frustrated commentary on the dissolution of the forms of life associated with industrial society or as a consequence of experiences of marginalization, but as an obvious socialist legacy. The official recognition of industrial workers in party ideology as well as the extensive efforts by the shipyards to provide an infrastructure for communal life (partly to compensate for a lack of state welfare facilities) clearly left its mark on the self-image of the workforce. The workers' memories symbolize the scepticism towards social individualization and a longing for an affective community centred on meaningful work with a tangible product.[16] In fact, most workers found employment elsewhere, but even those that remained in the industry described feelings of alienation and a lack of emotional connection with the shipyard. This is even more true for those who found jobs in the service sector – seasonal tourism, for example – which is unanimously said to create less community and less meaning.

We should not interpret these memories of a past filled with meaningful work merely through the prism of "restorative" nostalgia, that is, as an emotional desire to return to days long gone. For nostalgia can have a reflective dimension, it can be a way of dealing with losses, of accepting the past as past, and of devising a plausible plan for the future.[17] At first, our interviews do indeed seem to be in line with most of the anthropological literature on post-socialist workers, which emphasizes the multiple effects of loss of status on workers' consciousness and memories. Not surprisingly, most of the stories our interviewees told about the shipyards were imbued with nostalgia, as they had already lost their jobs, retired, or were soon to be laid off. Their descriptions contained a universal motif of decline, and the bankruptcy of the two shipyards lent additional credibility to such narratives. After all, we conducted our interviews at a time when the closure of the shipyard was only a few years back (Gdynia) or when it was in its final throes (Pula), meaning the wounds in the self-confidence of the proud workers were still very fresh.

In the course of the project, our doubts about this narrative grew and we asked ourselves whether we as interviewers had perhaps reinforced this pervasive culture of lament ourselves by way of our questions and curiosity. "Ordinary" people often suffer losses in times of far-reaching socio-economic change, they glorify their past and rant about "those upstairs" – relatively independent of their personal circumstances. These narratives can be formulaic and generic, based on media reports, or stories told by acquaintances. We therefore remain fundamentally agnostic to this rhetoric of decline and loss of control, not wishing to simply regurgitate what has been said without any element of criticism. As interviewers and participant observers, however, we can by no means deny that such processes have taken place; indeed, we are also familiar with them from "conventional" archival sources and sociological studies that have been produced with greater temporal proximity. We have therefore primarily traced the specific social practices and experiences and found that there are some similarities with the life stories of industrial workers who had experienced the much-discussed structural change in the West in the 1970s and 1980s.[18] What is therefore long overdue is a systematic comparison of the experiences of the capitalist and (post-)socialist paths to post-Fordism.

In both the West and the East, ubiquitous but very much time-specific narratives of catastrophe often conceal considerable creativity "from below." This is where the aforementioned concept of *habitability*, developed by social anthropologist Jeremy Morris to analyse the changes in a post-socialist monocity in Russia – a place that experienced social devastation on a much larger scale than Gdynia or Pula – is useful. Habitability, says Jeremy Morris, "arises out of 'small agency' that is locally and socially embedded."[19] People do everything they can to make their environment liveable – even if they conjure up images of decline when recounting their experiences, in most cases they use the opportunities that are present to make the best of things. People's positive relationship with work, the community relations inherited from socialism, internalized routines of improvisation, and the often hard-learned lesson that change is inevitable are socio-cultural resources on which future-oriented life strategies can be built. In many ways, the experiences of late socialism and early post-socialism equipped people with resources such as a talent for improvisation, dense social networks, as well as a willingness to break the rules that allowed them to build new lives on the (metaphorical) ruins of industrial modernity.

Unlike other work on post-socialism, our research shows that from the workers' perspective the period after 1989 cannot simply be described as a slow decline. For if firms and industrial lifeworlds can survive beyond the political upheaval, continuing to exist in 2009 or 2018, despite all the difficulties, there must have been a sufficiently large reservoir of social resilience. Workers contributed to this with their agency; at the same time the experiences they encountered since the 1980s offered them an easy explanatory framework in which to place "us" against "them" or "the establishment" – an attitude that populist politics served to reinforce.[20]

The workers' lives were also subject to a kind of "creative destruction" on a personal level, and yet they had the opportunity to interpret and – in part at least – shape the processes in their own way. For us, this was the deeper purpose of oral history, an important methodology for social historians that often unearths more insights into the dynamics of social relations and economic changes than written sources or performative acts such as balance sheet press conferences with seemingly objective indicators. Nevertheless, the latter are indispensable for writing corporate history in the conventional sense, because numbers not only feature constantly in the discourses of management, but also reflect the forces of the market. This perspective alone, however, would be too shallow, since an industrial enterprise is more than a production site equipped with machinery and also more than the difference between inputs and outputs. It is a world of meaning and relationships, a "business" in which people operate and cooperate, have experiences, develop visions, articulate protest – to put it in a nutshell, where history is made.

The Temporality of Transformation

The narratives we collected from workers, engineers, managers, experts, and local decision-makers may differ in their moral and political assessments of the situation. There is one point they are consistent on, however:

The transformation implied a fundamental change that affected not only the workplace, but also community structures and ultimately people's sense of identity. What the people we interviewed do not agree on is the chronology; the critical crossroads are difficult to pinpoint through interviews, as the interviewees tended not to put a date to the events they relayed. Such emic perspectives all but invite us to challenge the importance of 1989 (or 1990 in the case of Yugoslavia), i.e. the end of one-party communist rule, as a definitive caesura demarcating the before and after. Viewing transformation as structural change does not mean that individual political events do not matter, and it would be wrong to ignore the relevance of the end of communist rule for our story and the history of the region as a whole. On the contrary, this was a momentous event that strongly shaped the vector and speed of transformation. The change of government, however, neither triggered the transformation nor predetermined its outcome. And depending on the subject area, there are other timelines as well as factors that helped shape the process of transformation both before and after the upheaval.[21]

Identifying the beginning and the end of a complex and comprehensive process such as the transformation of this region is notoriously difficult – not only for those who lived through and helped shape it themselves, but even for historians, who by their very nature always have the benefit of hindsight. For this vantage point has its limits, since history never ends, the fundamental nature of history is change, and individual social spheres change at different speeds, leaving us unable to discern some of the underlying structures. The distinctive nature of transformation lies in the synchronicity of fundamental change across virtually every major social sphere. Multiple separate developments come together in one moment of upheaval: economic structures, social relations, political organization, international order, value systems, and norms were all transformed simultaneously in one consolidated process driven by interdependent and mutually reinforcing dynamics, albeit with often unclear directions of causality.[22] In this respect, one could interpret the transformation as a "more structural" change (if there were such a thing as a comparative of structural), i.e. change that encroached on the economy, politics, and the entire social order more strongly than the structural change that Western Europe has experienced since the 1970s.

A distinctive feature of the transformation is the fact that the socio-economic change was managed by political systems that had only just become democracies; many states in the region had also just gained their independence. Given the high level of uncertainty and the novelty of the situation in the early 1990s, policymakers and government institutions were simply overwhelmed by the challenges they faced. The visible hand of the state no longer provided unwavering support – and in a period characterized by the ideological hegemony of neoliberalism, it seemed out of place anyway.

Transformation can therefore be understood as a temporal, but also spatio-temporal compression of change, the outcome of which was by no means certain. The term transformation refers to a time when changes in different areas flowed into each other to such an extent that within just a few years' many people barely recognized what had once been a familiar environment to them. Unlike the terms revolution and turning point, the concept of the "saddle period" (*Sattelzeit*), coined by Koselleck, allows for the insight that even highly compressed change takes time. The saddle period is a phase of compressed fluidity in which old patterns are questioned and new structures emerge, in which actors are under stress, new futures are conceivable but not necessarily achievable, and horizons of expectation and spaces of experience – according to Koselleck – diverge before finding their way back to one another. Transformation serves as a link between the socialist version of modernity and an as yet undefined era. We do not yet know what term or designation will one day prove fitting for the period after the transformation. Regardless, from today's perspective, the terms "neoliberal" and "post-socialist" do not seem to refer so much to a new post-transformation phase as to the final stage of transformation before a new socio-economic model emerges.

When did those processes that led to what is now called "transformation" begin? We firmly believe that "Western" structural change and "Eastern" transformation started at the same time. In our case studies – two shipyards and the Eastern European shipbuilding industry as a whole – it is clear that the slump in demand for ships after the oil price shock was the catalyst for a structural reconfiguration.[23] The shipbuilders and governments of Poland and Yugoslavia found themselves facing the challenge of adjusting to this new situation, one over which they had little influence. Decision-makers were all too hesitant in their response, resulting in massive costs for the state and increasingly undermining its ability to act. Instead of efficient crisis management, they relied on outdated patterns of action that ultimately made the need for reform all the more urgent. The legitimacy of the communist system, which was based on the state's claim to absolute agency, suffered as a result. The reactions to the first oil price shock on the part of the government and companies, as well as the unintended consequences these reactions brought in their wake, set processes in motion that initially delayed structural change only to result in major transformation of an entirely different quality. What this teaches us is that even policies that are conservative in the literal sense of the word can lead to profound changes that then elude political control much more than before.[24] At the same time, the workers' protests in Poland in the late 1970s and the growing frustration with workers' self-management in Yugoslavia suggested that a more comprehensive change in political values and norms was in the offing; even "from below" the system was beginning to disintegrate.

Neither Poland nor Yugoslavia were able to implement policies that would have helped their shipyards remain competitive despite the contraction of the world market at the time. Downsizing was politically out of the question; on the contrary, capacities were expanded in the 1970s and 1980s. The innate logic of autarky of state socialist economic institutions was ill-suited to the foreign trade strategies of the currency-hungry communist governments. Generous loans from the West and the focus on exports served only to increase the tensions between domestic institutions and the expectations of foreign business partners. In another context, political scientists John Campbell and John Hall refer to what they call the "paradox of vulnerability," when particularly export-oriented countries (their comparative study uses the examples of Switzerland and Ireland), owing to what they perceive as their own vulnerability to global economic cycles, pull together in a collective effort to (successfully) build institutions that help them cushion external economic shocks; and that consequently make them particularly resilient to crises.[25] Poland and Yugoslavia obviously failed to achieve this paradoxical result. Instead, their vulnerability (e.g., vis-à-vis Western creditors) resulted from the communist regimes' refusal to face reality, which served only to make the post-1989 reforms all the more painful, intensifying the delayed structural change and turning it into a transformation.

Our case studies provide rich material illustrating Besnik Pula's assumption that it was the ever-increasing international integration that made socialist economies vulnerable.[26] While authors like Pula and Vladimir Unkovski-Korica[27] drew attention to this divergence by examining economic policymaking, we have turned their findings on their head: our work is one of the first studies on how the engagement of socialist enterprises in international markets affected the relations on their factory floors. From the point of view of the two shipyards, participation in international markets led to friction that neither their management nor the political system as a whole could absorb. Indeed, the governments in the two countries did not even manage to establish a subsidy model for their shipyards that would have been on a par with that in the West. The demands of global competitiveness, on the one hand, and the commitment to expanded social reproduction, on the other, ultimately proved incompatible, especially as the capacity of the state to provide continual assistance diminished as there were simply too many budget holes to plug. Both Poland and Yugoslavia had been heavily indebted to the West since the 1970s and slipped into a situation of insolvency in the 1980s. The shipyards were part of, in fact in many ways one of the causes of these macroeconomic problems. As the director of Uljanik once commented dryly, his shipyard was saved from closure only by the inherent reluctance of socialist economies to let firms exit the market.[28]

Thus, if, for once, we do not look at postwar history from the perspective of its supposed end in 1989, what we will come to recognize is that late socialism can in fact be seen as both the peak of "actually existing socialism" and, in terms of increasing integration into the world economy – along with the difficulties that ensued –, also the first phase of post-socialism. Under communist rule, this meant that the "basis" and the "superstructure" became increasingly incongruent, which, as any good Marxist will know, is the precondition for the emergence of a truly new historical formation.

If transformation began in the 1970s, how should we interpret the 1990s, which is usually considered the decade of transformation or (systematic) transition? In the case of the two shipyards we show a high degree of continuity with the 1980s. Crisis, "muddling through," and the struggle for survival continued, exacerbated by the deep economic recession in the two countries in the first "transition" years – and in Croatia by the war. What changed – faster in Poland than in Croatia – was the ideological framework and the economic institutions. Deregulation, liberalization, and privatization were now top of the political agenda, at least going by the speech acts of government representatives. At the company level, however, the direction of reform was often less clear. As pointed out by Appel and Orenstein, the neoliberal reform rhetoric that came from governments very much served as an instrument of "competitive signaling."[29] Through their impassioned commitment to the principles of neoliberalism and competitiveness, national governments hoped to attract foreign investors and bring their countries closer to EU membership.

However, speech acts alone will not simply erase firmly entrenched social and institutional practices. Our two case studies both reveal significant discrepancies between the macropolitical conditions and microeconomic business practices, even though both shipyards continued to be state owned for a number of years, meaning the respective governments could have directly intervened had they been inclined to. The institutional set-up that had developed in the 1980s proved to be remarkably resilient and although the ruling parties in Warsaw and Zagreb were all for privatization, they failed to follow it through. The two governments talked the competitiveness talk, even offering the requisite support to stop the shipyards from going under, but when it came to it, they barely invested anything in technical modernization. The state aid provided was thus nowhere near enough to put a process of successful restructuring in motion, something that would also have required some form of strategic direction. At the same time, the two governments also lacked the mettle to let the shipyards go bankrupt, even though this would have been the only sensible move, financially speaking. Despite the political rhetoric of privatization and marketization, the governments did not really seem to know what to do with the property they owned. They were caught in the trap of their own neoliberal rhetoric, finding themselves in something of a conceptual impasse when it came to acting like an owner of property they considered inappropriate to have in the first place.

The shipyard management were probably the most likely to try and break out of this ideological void. The managers – the new term used to describe the company directors from the 1990s on – attempted to adapt the shipyards to the new market conditions and modernize their manufacturing technologies. At the same time, when it came to maintaining relations with the government, the managers were – just as they had been under socialism – almost like "envoys" for their companies. In their efforts to tap into the government resources that were so essential for their survival, they used their tried and tested talent for presenting the shipyard's interests as national interests, drawing on the symbolic and social capital they had accrued under state socialism. The reduction in the workforce was more a result of natural shrinkage – caused by job changes and retirement – than targeted measures. The separation of areas that were not part of the shipyards' core business was also frequently more of an emergency measure to plug gaps in the budget than part of a coherent corporate strategy. Similarly, actions taken by interest groups, such as managers, employees, trade unionists, and political decision-makers, tended to be reactions to day-to-day concerns rather than an expression of longer-term planning. How could social and economic action, including on the micro level, be guided by a long-term strategy when the institutional conditions themselves were constantly changing?

With the legacy of socialist holism continuing to strongly influence the stakeholders, for a long time there was no consensus on what the core business of the shipyards actually was. Was the primary aim of the company to make money from selling ships or was it to create jobs and preserve community structures? In the 1990s, as our multiperspective approach has shown, the actors involved all came up with different answers to this question. And, the closer to the shipyard a particular group of stakeholders were, the more they leaned towards expanded reproduction as the main justification for the company's existence. More to the point, this also gave them a solid reason to make demands of the state – they were, after all, providing a service for the community that the government would otherwise have had to take care of itself.

The 1990s were thus more of a transition phase where the individual interest groups pursued very different objectives. The problem, however, is that this failed to bring about a clearly defined process of transformation, with government reformers signalling to international investors that their countries were opening up, while other stakeholders sought to develop a key "national" industry. The opposing paths of privatization that the two shipyards took are pertinent examples of this apparent contradiction. In both shipyards, as well as in other sectors of the economy such as banking, the state continued to play a bigger role than neoliberal buzz words such as "shock therapy" would suggest. The rhetoric used by the shipyards to legitimize their existence and justify the demands they made of the government most certainly still had to be adapted to the new ideological realities – the hegemony of neoliberalism, and in Croatia's case, ethnic nationalism. Scratching the rhetorical surface, however, reveals a wide array of different practices. Mimicry had already been a widespread practice under state socialism and now implied that the declarative commitment to reform did not necessarily have be accompanied by a belief in the ideological principles at the core of these reforms.

The developments that followed the first pioneering reform initiatives of the 1990s also show how differently the concept of privatization can be defined and operationalized. A prime example of this is the purchase of Uljanik by its employees in 2012. Neoliberalism served as no more than a surface ideology that provided a general reform framework (or language of reform) but when it came to the key sectors of the economy it often had no specific guidelines to offer. The shipyards' response to the situation was to permanently perform transformation.[30] They adapted their organizational toolbox to the new political and ideological conditions, while attempting to maintain a high degree of continuity within the company at the same time. This strategy had already served the shipyards well in the crisis of the 1970s.

After EU accession, however, the strategy was no longer effective. Did this signify the end of the major, long-lasting transformation? Or should the global financial crisis, which coincided with the closure of Gdynia Shipyard, be considered its endpoint? As Joseph Stiglitz and other economists have shown, and indeed criticized, after the 2008/9 stock market crash, the neoliberal order initially proved to be remarkably resilient, showing increased signs of dissolution only over the next decade.[31] In Poland and other post-communist countries, state dirigisme had already enjoyed a renaissance a few years earlier because neoliberal reform policy was not well received by the electorate, provoking populist backlash. The PiS party, which was back in power from 2015 to 2023, invoked not only "traditional values" and the Polish nation, but also a strong state as a direct expression of the will of the people. The PiS-led government's social and economic policy broke with the neoliberal zeitgeist, on the one hand by reversing what had previously been highly flexible employment contract options and, on the other, by introducing new social benefits as well as a higher minimum wage, albeit without calling the capitalist foundation of the Polish economy into question.[32]

As pointed out by both David Ost and Donald Kalb, for instance, PiS enjoyed great popularity among the workers, and indeed still does (much like the "blue-collar" workers in the United States and other countries such as Austria increasingly voted for right-wing populist parties).[33] PiS was able to fill the political vacuum left behind by the Polish Left when it committed to a neoliberal reform policy and no longer even attempted to see the workers as a group. What PiS did was to promise to take steps to repair the material and symbolic damage that the workers believed to be so terribly unfair. A new, just community – of "real" Poles – would emerge, one in which people no longer had to emigrate to Western Europe in order to become prosperous. As David Ost points out, this political message resonates in particular with the full-time "blue-collar" employees in industries with a comparatively high level of trade union membership – and it is this group that makes up PiS' "most loyal" base among the working class.[34] Now, as we near the end of the transformation, it appears that the erstwhile working class have finally changed their political alignment.

The question is whether this also includes "our" shipyard workers in Gdynia. We cannot be entirely sure, but perhaps not: Gdynia remains a stronghold of Poland's liberal parties. At the last Sejm elections in 2023, the liberal party alliance, led by the PO, won 47 per cent of the votes in the city, while PiS and the other parties on their list received only 20 per cent.[35] The PO candidate for the Senate, the second chamber of the Polish parliament, trounced the PiS candidate for Gdynia by 76 per cent to 24 per cent.[36] In the second round of the 2020 presidential elections, the (successful) PiS candidate Andrzej Duda was trailing his liberal challenger Rafał

Trzaskowski in Gdynia by 40 percentage points. Croatia, for its part, has not yet seen a strong right-wing populist backlash, with some of the old industrial strongholds – including Pula – primarily voting for the left-leaning parties. During the parliamentary elections in 2020, the centre-left alliance led by the Social Democrats secured more than 50 per cent of the vote in Pula, in some precincts even hitting over 60 per cent in electoral districts where a high percentage of former Uljanik workers cast their ballot.[37] Just as we called for acceptance when it came to the ambivalence of the economic results of transformation, the same applies to the political sphere. Right-wing populism can, but does not necessarily have to be the political response of a working class frustrated by their experience of loss – indeed, perhaps the working class are not even universally frustrated after all. There are other factors besides occupation that are no less important when it comes to political preferences – as is patently obvious if you look at Poland's or indeed Croatia's electoral map.

After Neoliberal Hegemony

In our reflections on the temporality of transformation, we were unable to determine an endpoint. Perhaps it was marked by the Covid-19 pandemic, in the wake of which the principles of neoliberal economic policy that had applied for the last 40 years were cast aside. Whatever the case, the awareness that dependence on East Asia, particularly in shipbuilding and maritime transport, has its drawbacks appears difficult to translate into practical policy let alone into the revival of European production sites. Or does it?

When it comes to the question of whether there might have been alternatives to a neoliberal post-socialist era, some of the shipyards' workers and observers at the time had a very clear answer, putting Brussels directly in the firing line: "If we weren't members of the EU-27, the trade unionists would come to Warsaw and the government would be forced to increase our pay. And that would give us four or five years of peace and quiet."[38] This is how a journalist for the national daily *Gazeta Wyborcza* viewed the situation in a 2008 article that addressed how the traditional negotiating patterns between companies and the Polish government had been unsettled by the European Commission's interference. Our study undoubtedly confirms that EU accession has been a frequently underestimated caesura in the transformation period. In 2004, 2007, and 2013 – the big rounds of EU enlargement in Eastern Europe – the comprehensive realignment process for the political economy and its normative and ideological foundation reached its first major endpoint.

This also heralded the start of the final phase of the major, long-lasting transformation. Hilary Appel and Mitchell Orenstein contend that, when it came to the neoliberal restructuring of the economy and making headway with the neoliberal project, the EU took over the baton from IMF and World Bank. "The EU helped to complete the unfinished business of privatization and liberalization."[39] The accession candidates had to accept the entire body of European Union law (*acquis communautaire*), including all its rules on state aid. This was to help the EU achieve its ambitious goal of becoming the world's most competitive economic area. The nation states and their actors were left with very little room for manoeuvre.[40]

Since the completion of structural change in Western Europe, the EU has fundamentally revised the position of state actors vis-à-vis businesses. The role of national governments has been transformed *nolens volens* from one of protector and reinsurer to that of an executive body tasked with enforcing competition law, with the additional aim of fostering the global competitiveness of European companies. Our shipbuilding example clearly shows just how much the rules of the economic policy game changed after EU accession. Aleksandra Sznajder Lee and Vera Trappmann came to very similar conclusions for the steel industry (albeit from a different perspective and, to some extent, with different outcomes, as the resilience of Eastern Europe's steel industry demonstrates).[41] In shipbuilding, EU legislation forced the accession countries, including Poland and Croatia, to abandon their long-established practices of providing direct and indirect support for the shipyards. In so doing, the EU rules essentially put a permanent stop to what János Kornai once dubbed "soft budget constraints," which, as we have shown, had been continued for a surprisingly long time after 1989.

The nerve-wracking EU accession negotiations already gave the world a foretaste of the new era of shipbuilding that was to come. As we have described, the EU Commission raised questions about the state subsidies that Gdynia had received, while Croatia, for its part, was forced to swiftly conclude the privatization and restructuring of its shipbuilding sector. It then dawned on the stakeholders that the difficult decisions they had avoided for so long could not be put off any longer. The Eurocrats at the Commission turned a deaf ear to sentimental stories about the symbolic meaning of shipbuilding and the pride associated with past export successes. Managers and trade unionists were stripped of the instruments of legitimation with which they had so far successfully acquired state aid and won over the local public. At the same time, the arrival of this supranational actor permanently changed the political calculus for the national governments, as they could now shift the blame for the demise of the shipyards onto Brussels. It is therefore no coincidence that our shipyards in Gdynia and Pula both continued to exist for the same amount of time after EU accession. Each survived for exactly five years. Poland joined the EU in 2004 and Croatia became a member in 2013, Stocznia Gdynia went bankrupt in 2009, Uljanik in 2018.

The transformation, and with it post-socialism, seem to have come to an end – so, what comes next (to paraphrase social anthropologist Katherine Verdery, who asked this very question in reference to socialism)?[42] Historians are not good at forecasting. If we had written our book a decade earlier, our conclusion might have featured a somewhat optimistic outlook for the post-socialist shipyard on the Adriatic. At the time, it seemed as though Uljanik had developed its own niche that would enable it to stand strong in the face of global competition. In Gdynia, in contrast, there was a pungent smell of scorched earth – both politically and economically speaking. The assumptions we made at the start of the project taught us to be very cautious about making firm predictions regarding the future – the shelf-life of such predictions would certainly be shorter than that of the ships, which are after all the real protagonists in our story.

One of the main reasons for this is the potential for dramatic events that render even the most rigorous prediction and the very best business strategy obsolete. The only constant in the story seems to be its sense of irony. This became apparent yet again when, during the Covid-19 pandemic, gigantic vessels were moored in the waters of the Uljanik shipyard. Four months after the *Willem van Rubroeck* dredger was transported from Pula to Gdańsk, the *Ibn Battuta*, a vessel of the same type, owned by the same company, docked in the Bay of Pula. Measuring 138 metres in length, the *Ibn Battuta*, named after a medieval Muslim scholar and traveller, is similarly imposing. The dredger was built by Uljanik ten years previously as a prototype for a series of ships commissioned by Luxemburg-based company Jan de Nul, which specializes in maritime construction work. An article in *Glas Istre* on the arrival of the dredger vessel commented that *Ibn Battuta* was an erstwhile "harbinger of a new future, which unfortunately never arrived."[43]

Alongside the *Leif Eriksson*, a second big dredger ordered by Jan de Nul, as well as a few other specialized ships from the same company, the *Ibn Battuta* has now been anchored since August 2020 in the very same place that the *Willem van Rubroek* once awaited its completion. These ships were brought to Pula because, amid the Covid-19 pandemic, their owner considered the city to be a safe harbour in the literal sense of the word and, with Uljanik having gone bankrupt, the shipyard

Figure 6.1. The dredger *Ibn Battuta* in the shipyard in Pula, 2021
Source: Photograph by Ulf Brunnbauer.

had space for such massive vessels (figure 6.1). Having dramatically reduced its operations during the pandemic, Jan de Nul was forced to temporarily decommission the two ships. Thus, the shipyard had, at least temporarily, found itself a new role – no longer a production site, it now served as a convenient parking lot for the phase between the old and the new "order" in the shipbuilding industry and potentially even the global economy.

Postscript

In May 2021, at the 12th National Maritime Conference in Rostock, the then German economics minister, Peter Altmaier (CDU), lamented, "In some areas outside Europe, we're seeing a competitive landscape that calls for action on the part of the European Union."[1] What Altmaier was referring to here was the shipbuilding industry, especially in countries "that very clearly and at times quite aggressively are trying to monopolize or dominate parts of civic shipbuilding," which is why "we cannot stand by and let this happen." And while he did not explicitly name China, it was clear whom he had in mind. The interruption of supply chains during the Covid-19 pandemic provoked a debate in Germany and the EU that ten years earlier, even during the euro crisis, had seemed out of the question: forward-looking industrial policy involving state intervention in the market ceased to be anathema, not only to protect certain sectors but also to support the emergence of what have been dubbed national or European "champions." This is not a new strategy, however. In fact, the EU has been successfully pursuing it with Airbus for decades.

In any event, the European shipbuilding industry sensed its opportunity, especially because all the talk about supply chain security provided it with a new argument to justify its position. In late April 2021, the Shipyards' & Maritime Equipment Association of Europe (SEA Europe) sent an open letter to the president of the European Commission, Ursula von der Leyen, and the executive vice-presidents for competition and the economy, respectively, Margrethe Vestager and Valdis Dombrovskis.[2] The industry representatives bemoaned that for "decades, the European Commission [has been] fully aware that European shipyards suffer from a unique legal gap in trade defence instruments." While their Asian competitors continued to profit from state assistance, the EU aid regime was becoming ever more restrictive, all in the name of competition. "Europe's market share declined from 45% in the 80ies to 5% today." It is about time that the EU tackle unfair global competition and protect the European shipyards. At the end of the day, shipbuilding is a strategic industry that provides innumerable jobs. "The time for action is now," SEA Europe stressed in its letter. The pandemic has been instrumental in exposing all the major risks inherent in Europe's dependence on other countries. The industry body linked its call for EU action on behalf of European shipyards with EU policy priorities (in a way that was not dissimilar to the rhetoric of legitimation used by the two shipyards): "Without its own shipyards, Europe will become entirely dependent on Asian shipbuilding to build or retrofit zero-emission ships – in line with the European Green Deal – that the EU needs to protect its defence and maritime borders and to access its seas, trade and Blue Economy."[3] And these claims were made at a time when, due to the upheavals caused by the pandemic, the global shipping industry was suddenly suffering from a lack of capacity in the face of record orders for new container ships.[4]

The economic recession caused by the Covid lockdowns along with massive pandemic-related problems with the delivery of medical supplies and other basic commodities, largely manufactured in China, served as something of a wake-up call for Brussels, Berlin, and other European capitals. The EU feared they would lose their technological and economic sovereignty – and sovereignty is a powerful word that is capable of legitimizing state intervention. These concerns even caused politicians who had long been committed to austerity and free trade, such as Peter Altmaier, to publicly consider market intervention in support of domestic industries. In February 2022, any remaining reservations about such interventions paled into insignificance as Russia launched its unprovoked attack on Ukraine, beginning a war that, as of summer 2024, when we finished the English version of our book, was still not over thanks to the heroic resistance of the Ukrainian people. The war made the importance of trade ships all the more evident, with the Russian Black Sea blockade and attacks preventing ships from calling at Ukrainian ports for months, hindering the transport of Ukrainian grain that is so crucial for ensuring global food security.

In response to the economic disruption caused by the war and high inflation caused by exploding energy prices, EU member states, with the blessing or even at the request of the Commission, were able to intervene in the market to a degree that just a few years ago would have been an anathema to the Eurocrats in Brussels. The relic of Keynesianism was dug out of the putative dustbin of history and dusted off, while neoliberalism suddenly seemed terribly outdated. The United States, once the vanguard of neoliberalism, has actually turned out under the Biden administration to be one of the more radical promoters of this shift towards industrial policy characterized by massive government subsidies and intervention in free trade – even prompting the EU to take similar steps. For the Gdynia and Uljanik shipyards, this paradigm shift comes too late, just as it does for most shipbuilders in Germany, France, and other EU states. But perhaps, politically speaking, its arrival is just in time for a Europe in which many, not least in Croatia and Poland, see themselves as victims of European integration and expect government economic policy to provide so much more than a mere guarantee of fair internal competition.

Notes

1. Weathering the Storms of Transformation

1 *S.A.* is the Polish abbreviation for *spółka akcyjna* (joint-stock company).
2 Raphael, *Jenseits von Kohle und Stahl*. In many ways, we see our book as a temporal and spatial extension of this seminal work. Steel is still used at Crist S.A., of course, albeit in much smaller quantities than in the former Gdynia Shipyard.
3 See the OECD data on these indicators at "Regional Economy: Gross Domestic Product, Small Regions TL3," OECD.Stat, accessed 29 March 2024, https://stats.oecd.org/index.aspx?queryid=67051. The Polish data grouped Gdynia, Gdańsk, and Sopot together as the "Trójmiasto." However, Gdynia is certainly no poorer than its two neighbouring cities. The GDP per capita (purchasing power parity) of these cities in 2018 was 45,288 US dollars in the Trójmiasto, 45,016 US dollars in Rostock, and 43,775 US dollars in Bremerhaven (data retrieved May 2021).
4 On the debts of the Croatian shipyards at the time, see Bajo and Primorac, "Jesu li brodogradilišta."
5 Calic, "Beginning of the End."
6 Bartel, *Broken Promises*.
7 Murphy and Tenold, "Appendix 1."
8 See Berend, *From the Soviet Bloc*, 7–37.
9 Bartel, *Broken Promises*.
10 Other major investments during this period included the power plant in Bełchatów, the automobile factory in Tychy, the refinery and the new port in Gdańsk (Port Północny), and several new train and motorway connections.
11 On the notation of place names: in this book, we use English names for cities if they are in common use, such as Moscow or Warsaw. No political intent on our part should be inferred from this usage; many other languages besides English also use localized forms of place names.

12 Schierup, *Migration, Socialism*.
13 Sachs and Lipton, "Poland's Economic Reform."
14 Ghodsee and Orenstein, *Taking Stock of Shock*.
15 Hann, *Postsocialism*; Kideckel, *Getting By*; and esp. Ghodsee and Orenstein, *Taking Stock of Shock*, pt. 4.
16 See Polanyi, *Great Transformation*. The original version appeared in 1944 and was reprinted numerous times after the war. On Polanyi, see Dale, *Limits of the Market*; and Dale, *Life on the Left*.
17 See the particularly well-written work by Bohle and Greskovits, *Capitalist Diversity*.
18 From 2016 to 2021, the project was funded by the German Research Association and the Austrian Science Fund and jointly coordinated by the Leibniz Institute for East and Southeast European Studies (Regensburg) and the Institute of East European History at the University of Vienna. See for a Polanyian take on the post-1989s transformation also Ther, *How the West Lost the Peace*, xii–xvii, 2–9.
19 This has already been observed by political scientist David Ost in his important and, as it would later turn out, prophetic book *Defeat of Solidarity*.
20 According to calculations by the EU Commission, this was the sum of the subsidies later declared illegal. See European Commission, "Commission Decision of 6 November 2008," 26, item 232. On Croatia, see Bajo, Primorac, and Hanich, "Financial Performance," 9. For further details and data, see chapter 2.
21 Slobodian, *Globalists*.
22 On this as well as on the relationship between democracy and economic reforms, see the still relevant book by political scientist Orenstein, *Out of the Red*.
23 The data on the share of industrial production in GDP and total employment are taken from World Bank Open Data, accessed 28 March 2024, https://data.worldbank.org/.
24 Maddison, *World Economy*. On the catch-up after 1989, see Ther, *Europe since 1989*, 126–9, 200–16.
25 Ghodsee and Orenstein, *Taking Stock of Shock*.
26 Data from "GDP per capita in PPS," Eurostat, last updated 6 March 2024, https://doi.org/10.2908/TEC00114.
27 On this term, see, e.g., Stark, *Sense of Dissonance*.
28 One of the most important emblematic moments of the "Warsaw Consensus" was the opening of the first McDonald's restaurant in Warsaw by Jacek Kuroń. See "Big Mac nachodzi," *Polska Kronika Filmowa*, 17 June 1992, YouTube video, 2:34, https://www.youtube.com/watch?v=0wnA6QxuGEw.

29 As observed by political scientist Adam Przeworski as early as 1991 (Przeworski, *Democracy and the Market*). On the parallels between the Washington Consensus and the "Warsaw Consensus," see Ther, "Globale Hegemonie des Neoliberalismus."
30 On the collapse of Yugoslavia, see Woodward, *Balkan Tragedy*; and Sundhaussen, *Jugoslawien und seine Nachfolgestaaten*.
31 The telling title of their article, which addresses the Russian Federation following the invasion and annexation of Crimea, is "Normal Countries." See Shleifer and Treisman, "Normal Countries."
32 See chapter 2, as well as "Sanacija i restrukturiranje." The company received a debt remission worth 1.4 billion kuna (almost 200 million euros according to the exchange rate at the time), and 771 million kuna of new shares were issued as well.
33 Barrett, "Informal Networks."
34 See Appel and Orenstein, *From Triumph to Crisis*, 111–41.
35 See the list of liquidated enterprises in the Polish food industry and other sectors compiled by Karpiński et al., *Jak powstawały i jak upadały*, 323–6.
36 Mirjana Vermezović Ivanović, "Kako je uništena industrija u Istri," *Deutsche Welle*, 6 February 2020, https://www.dw.com/hr/kako-je-uni%C5%A1tena-industrija-u-istri/a-52275942.
37 On the further development of numerous privatized large companies and especially their work culture, see Wawrzyniak and Leyk, *Cięcia*.
38 From 1994 to 1996, Szlanta had served as director of the Polish Development Bank (Polski Bank Rozwoju). Having completed a degree in physics, he owed this position in part to his connections in the Unia Demokratyczna, a liberal Christian democratic reform party emerging from Solidarność and led by the first post-communist prime minister of Poland, Tadeusz Mazowiecki. An interesting aspect of Szlanta's career is that he was not replaced by the post-communists – presumably due to his success as head of the Polish Development Bank.
39 On the global development of shipbuilding and developments in various locations, see Varela, Murphy, and van der Linden, *Shipbuilding and Ship Repair Workers*.
40 See Berend, *From the Soviet Bloc*, 6–38. On the decline of shipbuilding in Western Europe, see Stråth, *Politics of De-industrialisation*.
41 See Keat, "Fallen Heroes."
42 See Monika Kosińska, "Janusz Szlanta nagrodzony: Najlepszy menedżer," *Dziennik Bałtycki*, 29 January 2001, 10.
43 See Agneza Urošević, "Do danas Uljanik isporučio 250 brodova," *Poslovni tjednik*, 8 December 2006, https://www.poslovni.hr/domace/do-danas-uljanik-isporucio-250-brodova-28597.

44 The lettering on the huge cranes of the dry dock was recently changed. It now reads "Crist S.A." instead of "Stocznia Gdynia S.A."
45 One of the first contributions to this discourse was the self-critical analysis by Król, *Byliśmy głupi*.
46 See the submissions for the competition "Rok 1989: koniec, przełom, początek …?" (1989: End, turning point, beginning …?). The KARTA centre's archive holds a total of 144 texts ("Komunikat Jury konkursu 'Historia Bliska,'" Fundacja Ośrodka KARTA, accessed 15 May 2024, https://karta.org.pl/aktualnosci/komunikat-jury-konkursu-historia-bliska-0). The learning plattform Uczyć się z historii, organized by KARTA, provides information on the competition and the awardees ("XIII edycja konkursu 'Historia Bliska,'" Uczyć się z historii, accessed 15 May 2024, https://uczycsiezhistorii.pl/konkurs/rok-1989-koniec-przelom-poczatek/). Ghodsee and Orenstein, in ch. 9 of *Taking Stock of Shock*, provide a succinct summary of public opinion surveys on transformation. Leaving substantial differences between countries aside, their evidence points to widespread dissatisfaction even where material living standards have improved since 1989.
47 On the decline of trust, see Ghodsee and Orenstein, *Taking Stock of Shock*, ch. 12.
48 See Musić, "Provincial, Proletarian, and Multinational."
49 Stanić, "'Jedan od najtežih dana u Uljaniku!'"
50 The ferry *Elektra* was a prestigious and successful project. It was, in fact, awarded the international "Ship of the Year" distinction in 2017. Several other shipbuilding-related companies are active on the former Stocznia Gdynia premises as well, including Energomontaż S.A. (EPG [website], accessed 13 March 2024, https://epgsa.com/epgen/), Gafako (Gafako [website], accessed 13 March 2024, https://www.gafako.pl), and the state-run repair yard Nauta (Shiprepair Yard Nauta [website], accessed 13 March 2024, https://nauta.pl).
51 See, e.g., Niethammer, von Plato, and Wierling, *Die volkseigene Erfahrung*; or the older study by Portelli, "What Makes Oral History Different." We mention these two authors and works in part because they reflect in depth on the theoretical problems of oral history.
52 See the chapter on Maurice Halbwachs in Niethammer, *Kollektive Identität*, 314–66.
53 For this older perspective of oral history, see, e.g., Thompson, *Voice of the Past*; and von Plato, "Oral History als Erfahrungswissenschaft." Later, however, the founding generation added to and partly revised its views.
54 See in detail Abrams, *Oral History Theory*.
55 The important role of the EU in the area of legal transformation is emphasized in Patel and Röhl, *Transformation durch Recht*.
56 Giddens, *Constitution of Society*.
57 On media history and the use of media sources, see, e.g., Bösch, *Mediengeschichte*.

58 On methodology and types of comparison, see, e.g., Siegrist, "Perspektiven der vergleichenden Geschichtswissenschaft."

2. Forever on the Verge of Going Under

1 "Predsjednik Hua Koa Feng posjetio Uljanik," *Mali Informator*, 31 August 1978, 1; and "Chairman Hua's Yugoslav Itinerary," *Peking Review* 21, no. 35 (1978): 9.
2 N. Trgovičić, "'Uljanik' će graditi pet brodova," *Glas Istre*, 3 August 1978, 3.
3 Colton and Huntzinger, *Shipbuilding in Recent Times*, 17.
4 Milan Pavlović, "Uljanik i 3. maj su jednako tehnološki zaostali," *Glas Istre*, 25 May 2019, https://www.glasistre.hr/pula/uljanik-i-3-maj-su-jednako-tehnoloski-zaostali-kad-su-bili-u-posjeti-kinezi-sumislili-da-obilaze-muzej-brodogradnje-589279.
5 Steidel, Daniel, and Yildira, *Shipbuilding Market Developments*.
6 Wegenschimmel, *Zombiewerften oder Hungerkünstler?*
7 Bitzer and von Hirschhausen, "Shipbuilding Industry," 33–4.
8 There is already substantial literature on the 1970s as a watershed moment in global history: see, e.g., Jarausch, *Das Ende der Zuversicht?*; Doering-Manteuffel and Raphael, *Nach dem Boom*; Hellema, *Global 1970s*; Ferguson, *Shock of the Global*; and Bartel, *Broken Promises*. Bartel highlights the importance and transformative consequences of the oil crisis, as well as the divergent responses of capitalist and socialist states to these shocks.
9 Pula, *Globalization*, 66.
10 Bartel, *Broken Promises*, 49–50.
11 See Raphael, *Jenseits von Kohle und Stahl*.
12 Pula, *Globalization*, 103.
13 Kotkin, *Uncivil Society*, 28–30.
14 Calic, "Beginning of the End."
15 Bartel, Broken Promises.
16 Van der Linden, Murphy, and Varela, "Introduction," 15.
17 Šesnić, "Posljednjih 60 godina Brodogradilišta Uljanik."
18 "'Uljanik' prvi izvoznik-proizvođač."
19 Murphy and Tenold, "Appendix 1," 667.
20 Murphy and Tenold, 667.
21 "Zapisnik od proširene sjednice Republičkog odbora sindikata radnika proizvodnje i prerade metala SR Hrvatske," Rijeka, 14 November 1975, HDA, f. 1398, kn. 1, p. 8.
22 Warlouzet, "Collapse of the French Shipyard," 861.
23 Slaven, *British Shipbuilding*, 212.
24 Jadranbrod, "Stanje, ekonomski položaj i problemi nezaposlenosti kapaciteta brodogradilišta," Zagreb, June 1980, HDA, f. 1398, p. 19; SRH, IVS (Executive Council of the Socialist Republic of Croatia), "Platforma za rješenje problema brodogradnje i brodarstva," 14 July 1977, HDA, f. 280, kut. 108, br. 979.

25 See Warlouzet, "Collapse of the French Shipyard."
26 Lammers, "Subventionen," 11; cf. Wolf, "Bremer Vulkan," 128.
27 European Commission, "Communication from the Commission to the Council."
28 Warlouzet, "Collapse of the French Shipyard," 861.
29 Murphy and Tenold, "Appendix 1," 672.
30 Murphy and Tenold, 672; and Slaven, *British Shipbuilding*, 214.
31 Slaven, *British Shipbuilding*, 214.
32 Murphy, "Shipbuilding and Ship Repair Industries," 61, 78, 103.
33 ECORYS, *Study on Competitiveness*, 41; Nam, *Building Ships*; and Bruno and Tenold, "South Korea's Ascent."
34 Van der Linden, Murphy, and Varela, "Introduction," 20; and Shin, "Evolution of Labour Relations."
35 Lammers, "Subventionen," 6, table 3.
36 Collins and Grubb, *China's Dynamic Shipbuilding Industry*.
37 Lipanović and Kašper, "Jučer, danas, sutra," 6.
38 See Jessop, "Spatial Fixes."
39 Ethnologist Andrea Matošević analysed both documentaries in his article "Tehnička događajnica." The *Berge Istra* sank in 1975, resulting in thirty fatalities; its sister ship, the *Berge Vanga*, was to suffer the same fate in 1979.
40 David Fištrović, "'Mamuti' budućnost 'Uljanika,'" *Glas Istre*, 14–15 July 1973, 3.
41 M.U., "'Uljanik' – teškoće zbog duga," *Vjesnik*, 2 July 1972; and David Fištrović, "'Mamut' pokrio mamutski gubitak," *Novi list*, 4 September 1974.
42 M. Urošević, "Dva porinuća jednog broda – 'Uljanikova' ideja," *Vjesnik*, 1 September 1969; and Brigita Peko, "Mamut – brodovi na dva dijela," *VUS*, 7 August 1974.
43 Lampe, *Yugoslavia as History*, 310.
44 Martinčević, "Samoupravna normativna djelatnost."
45 Djeković, "Außenwirtschaftssystem und Außenwirtschaftsreformen," 71.
46 "Uspjesi na svim poljima," *Brodograditelj*, 26 December 1974, 5.
47 R.R., "Uljanik u potrazi za poslom," *Glas Istre*, 24–5 December 1976, 14.
48 Dino Bedrina and Ljiljana Vojnić, "Rezultati ispitivanja javnog mnjenja među radnicima 'Uljanika,'" *Uljanik*, 12 July 1982, 15.
49 Schierup, *Migration, Socialism*, 234.
50 Kornai, "Soft Budget Constraint," 10.
51 Kornai, 10.
52 Sajatović, "Prenosimo," 178.
53 M. Barak, "Brodogradnji voda do grla," *Glas Istre*, 26 October 1976, 3; and "Nezaposlena brodogradnja," *Glas Istre*, 24 December 1976, 3.
54 "Zapisnik sa sjednice Republičkog odbora," Split (Brodosplit), 16 October 1976, HDA, f. 1398, kn. 2 (1976), p. 10.
55 "Wyniki produkcjno-ekonomiczyne 1988–1990," 26 April 1989, APG, 130, p. 8.

56 "Analiza działalności stoczni za 1978 rok," March 1979, APG, 130/2358, p. 10.
57 "Analiza działalności stoczni za 1978 rok," p. 12.
58 IVS, Letter to Sekr. financija SRH, 23 July 1976, HDA, f. 280, kut 57, br. 875.
59 CIA Directorate of Intelligence, *Yugoslavia*, 3.
60 Stanić, "'Jedan od najtežih dana u Uljaniku!'"
61 Grala, *Reformy gospodarcze w PRL*.
62 Miroslav Kubacki, "Jakie są szanse 'Komuny'?," *Głos Stoczniowca*, 26 September 1980, 10.
63 Tadeusz Daszyński, "Rozmowa z dyrektorem generalnym ZPO mgr. inż. Mieczysławem Tokarzem," *Głos Stoczniowca*, 2 January 1981, 4.
64 Janusz Wikowski, "Jednością silni," *Głos Stoczniowca*, 11 December 1983, 1.
65 "Program modernizacji i rozwoju przemysłu okrętowego na lata 1983–1985 i do roku 1990," March 1983, APG, 2371/13, p. 5.
66 Tadeusz Baumberger, "Sukcesem jest spoistość branży. Rozmowa z dyrektorem Zrzeszenia Przedsiębiorstw Przemysłu Okrętowego inż. Bolesławem Ślepowrońskim," *Głos Stoczniowca*, 26 February 1984, 4.
67 Documents in the archive of the Executive Committee of the Assembly (Sabor) of the Socialist Republic of Croatia – i.e., its government – indicate the high frequency of such meetings but also their meagre results; see HDA, f. 280 (SIV).
68 Republički sekretarijat za energetiku, industriju i zanatstvo and Republički sekretarijat za pomorstvo, saobraćaj i veze, "Informacija o prolematici brodogradnje i brodarstva s prijedlogom mjera," Zagreb, January 1975, HDA, f. 280, kut. 33, br. 1140.
69 Republički sekretarijat za energetiku, industriju i zanatstvo and Republički sekretarijat za pomorstvo, saobraćaj i veze, "Informacija," pp. 2–3.
70 "Informacija o prolematici brodogradnje i brodarstva s prijedlogom mjera," HDA, f. 280, kut. 33, br. 1140, pp. 4–5; and IVS, "Stanje, ekonomski položaj i problemi nezaposlenosti kapaciteta brodogradilišta," June 1979, HDA, f. 280, kut. 174, br. 481, p. 2.
71 Jadranbrod, "Sadašnje stanje, ekonomski položaj i problemi razvoja brodogradnje," Zagreb, June 1980, HDA, f. 1398, kn. 6, p. 9.
72 Jadranbrod, "Sadašnje stanje," p. 6; and "Materijalni položaj jugoslavenske brodogradnje," 12.
73 Palairet, "Croatian Shipbuilding in Crisis," 762.
74 "Bilješka o vodjenim razgovorima članova delegacije Sabora SRH," 25 May 1979, HDA, f. 280, kut. 174, br. 481.
75 IVS, "Stanje, ekonomski položaj i problemi nezaposlenosti," June 1979, p. 1.
76 "Program rada za 1976. godinu," Zagreb, January 1976, HDA, f. 1398 (1976), p. 10.

77 Radna grupa Republičkog sekr. za energetiku, industriju i zanatstvo, Republičkog sekr. za pomorstvo, saobraćaj i veze, and Jadranbrod, "Informacija o provodjenju mjera iz 'Platforme za rješenje problema brodogradnja i brodarstva,'" 26 October 1977, HDA, f. 280, kut. 108, br. 979; and Jadranbrod, "Predmet: Prijedlog 'Društvenog dogovora o osiguravanju i usmjeravanju sredstava za plasman domaće opreme i brodova u zemlji,'" 16 December 1986, HDA, f. 280, kut. 337, br. 618.
78 G. Moravček, "Rekorder pred bankrotstvom," *Večernje novosti*, 31 October 1987.
79 Jadranbrod, "Traženje povoljnije kreditne politike za brodogradnju u 1980. godini," 11 December 1979, HDA, f. 280, kut. 181, br. 627–93.
80 Palairet, "Croatian Shipbuilding in Crisis," 783–4.
81 "Nacrt zakona o pretvaranju kratkoročnih kredita danih Jugoslavenskoj banci za međunarodnu ekonomsku suradnju dugoročne kredite za namjene dopunskog kreditiranja izvoza opreme, brodova i izvođenja investicijskih radova u inozemstvu," AS-715, February 1986, HDA, f. 280, kut. 329, br. 195.
82 "Nacrt zakona o pretvaranju kratkoročnih kredita," February 1986.
83 Markulinčić and Debeljuh, *Uljanik 1856–2006*, 58.
84 Sajatović, "Prenosimo," 179.
85 "Zapisnik sa sjednice Republičkog odbora," Split, 16 October 1976, HDA, f. 1398, kn. 2.
86 Jadranbrod, "Traženje povoljnije kreditne politike za brodogradnju u 1980. godini," 11 December 1979.
87 Silvana Jakus, "Kriza na svjetskom tržištu," *Borba*, 29 August 1979; the losses are detailed in "Stenografski zapisnik sa zajedničke (proširene) 10. sjednice," 1 July 1980, HDA, f. 1398, kn. 6.
88 Stopford, *Maritime Economics*.
89 Graber Majchrzak, "Gdańsk Shipyard," 381.
90 Bartel, *Broken Promises*, 155–6.
91 Babić and Primorac, "Yugoslav External Debt," 69.
92 CIA Directorate of Intelligence, *Yugoslavia*.
93 Kumar Lahiri, "Yugoslav Inflation and Money."
94 Bartel, *Broken Promises*, 204.
95 CIA Directorate of Intelligence, *Yugoslavia*.
96 Jarosław Mykowski, "Szwedzi są zadowoleni. Z Per Bomanem, głównym projektantem Stena Line AB rozmawia Jarosław Mykowski," *Głos Stoczniowca*, 13 September 1987, 1.
97 "Wystąpienie pokontrolne," 10 July 1986, ANIK, PCM-41008-9-86, p. 3.
98 "Wystąpienie pokontrolne," 10 July 1986, p. 1.
99 "Wystąpienie pokontrolne," 10 July 1986, p. 2.

100 "Uchwała Nr 42/84 Rady Ministrów z dnia 19 marca 1984 r. w sprawie udzielenia pomocy finansowej stoczniom produkcyjnym dla poprawy ich gospodarki oraz zapewnienia realizacji zadań gospodarczych w latach 1984–1985," AAN, 1757/2-136.
101 "Uchwała Nr 42/84."
102 Jadranbrod, "U vezi za ranije dostavljenim materijalima," 30 June 1981, HDA, f. 280, kut. 218, br. 516.
103 Palairet, "Croatian Shipbuilding in Crisis," 777.
104 Mijodrag Sajatović, "Karlo Radolović," *Start*, 31 May 1986; and S. Maršić, "Želja je da se stvore jake radne organizacije i jak SOUR," *Uljanik*, 1 July 1983.
105 Sajatović, "Karlo Radolović."
106 "'Uljanikova' suradnja s danskom tvrtkom 'DUE,'" *Mali Informator*, 11 December 1980, 1–4.
107 "Ukorak sa svijetom (1975–1986)," *Uljanik*, 5 December 1986; and Mario Buršić, "CAD/CAM instaliran u 'Uljaniku,'" *Uljanik*, 1 September 1985.
108 Boris Kalčić, "Godina velikih događaja," *Uljanik*, 1 January 1986.
109 Markulinčić and Debeljuh, *Uljanik 1856–2006*, 135.
110 "Osnovali vlastitu brodarsku kompaniju," *Politika*, 19 September 1988; and Zvonko Tarle, "Bankrot-brodovi," *Nedjeljna Dalmacija*, 19 October 1986. The company was privatized in 1994, dissolving all ownership links with the shipyard. In 2022, it changed its name to Alpha Adriatic to avoid any association with the bankrupted shipbuilder.
111 "Najprofitnji gospodarski sustav može biti – brodogradnje," *Privjredni Vjesnik*, 22 April 1992.
112 Ryšlavy, "Hrvatska brodogradnja danas," 107.
113 Mirko Urešević, "Posao za 1250 brodograditelja," *Vjesnik*, 25 November 1986.
114 A.M., "Za sva mora svijeta," *Vjesnik*, 10 December 1986.
115 V. Čamdžić, "'Rolls Roycevi' brodogradnje," *Glas Istre*, 21 January 1987.
116 Goran Drenjak and Tanja Marčela-Volarić, "Brodogradnja u 'Bermudskom trokutu,'" *Glas Istre*, 3–4 October 1987.
117 Mirko Urošević, "Neizvjesnost na navozima," *Vjesnik*, 14 October 1988.
118 Jadranbrod, "Predmet: Nenaplaćena potraživanja 'Uljanika' i '3. Maja' iz Sudana," 30 March 1987, HDA, f. 280, kut. 366, br. 743.
119 G. Moravček, "Rekorder pred bankrotstvom?!," *Večernje novosti*, 31 October 1987.
120 "Blokada, deblokada," *Uljanik*, 15 September 1989.
121 See data at World Trade Organization, *International Trade Statistics*, 27.
122 Stopford, "Current and Past Policies."
123 *Uljanik Brodogradilište Strateški plan, 1991–1996*, translation of the (unavailable) English original *Uljanik Shipyard Strategic Plan for 1991–1996*, PriceWaterhouse/IKO, May 1991, HFP Archive, no signature, p. 3.

124 *Uljanik Brodogradilište Strateški plan*, 16A–17A.
125 Stojčić, "Croatian Transition," 69.
126 See data in Bykova et al., *wiiw Handbook of Statistics*, table II/1.7; and Ther, *Neue Ordnung*, 90.
127 Keat, "Fallen Heroes," 212.
128 Keat, 212.
129 Jan Kreft, "Zakład do usług," *Dziennik Bałtycki*, 12 July 1995, 1.
130 Ostojić, "Novi vjetrovi," 152.
131 Keat, "Fallen Heroes," 218.
132 Keat, 219.
133 "Ranking Polityki. Największe polskie firmy," *Polityka*, no. 23 (1994): 18.
134 Jan Kraft, "Wzięliśmy sprawy w swoje ręce," *Dziennik Bałtycki*, 7–8 February 1998.
135 Keat, "Fallen Heroes," 220.
136 "Lista referencyjna," Stocznia Gdynia S.A., accessed 14 March 2024, http://www.stocznia.gdynia.pl/listarefer.php.
137 Kosińska, "Najlepszy menedżer," 10.
138 Kosińska, 10.
139 Colton and Huntzinger, *Shipbuilding in Recent Times*, 27.
140 Schönfelder, "Schmerzlose Stabilisierung?"
141 Milat and Galić, "Croatian Shipbuilding Industry," 15.
142 Mirko Urošević, "Brodovi na 'suhom,'" *Vjesnik*, 29 August 1992.
143 "Dostava odluke o pretvorbi i Programa vlaniščkog restrukturiranja," 6 November 1991, HFP Archive, *Elaborat*, no signature.
144 R.R., "Danas 'Uljanikovci' štrajkuju četiri sata jer nisu dobili plaće za rujan," *Glas Istre*, 19 October 1992, 8; and R. Radošević, "Štrajk zbog neimaštine," *Glas Istre*, 20 October 1992, 3.
145 T.U.I., "Plaće prekinule štrajk," *Glas Istre*, 22 October 1992, 1–2.
146 R.R., "Štrajk brodograditelja zbog malih plaća," *Glas Istre*, 15 April 1994, 16; and R.R., "Prekinut dvodnevni štrajk," *Glas Istre*, 16 April 1994, 15.
147 J.O., "Jučer je završeno prijavljivanje učenika za upis u prvi razred srednje škole. Brodogradnja ne pobuđuje interes," *Glas Istre*, 8 July 1993, 8.
148 Sladovljev, "Restrukturiranje hrvatske brodogradnje," 290.
149 Baučić, "Početak restrukturiranja."
150 Manuela Tašler, "Brodogradilišta idu na remont," *Vječernji list*, 22 June 1995.
151 Snježana Vujisić, "Odsukavanje 'škverova' financijskom polugom," *Vječernji list*, 25 October 1995.
152 R. Radošević, "Uzdah olakšanja, vjera u oživljavanje brodogradnje, ali i poveća doza skepse," *Novi list*, 28 October 1995.
153 Čuvalo, "Projekt – rekonstrukcija."
154 R. Radošević, "Stranci hoće hrvatska brodogradilišta, ali s čistim računima," *Novi list*, 14 March 1996.

155 Čuvalo, "Program restrukturiranja."
156 Manuela Tašler, "Brodogradilišta idu na remont," *Vječernji list*, 22 June 1995; and R. Radošević, "Višak 1.565 'Uljankovaca' – 500 otkaza," *Novi list*, 11 April 1996.
157 Markulinčić and Debeljuh, *Uljanik 1856–2006*, 154.
158 Republika Hrvatska, Državni ured za reviziju, "Izvješće o objavljenoj reviziji pretvorbe i privatizacije. Uljanik, Pula," Zagreb, July 2004, HFP Archive, no signature, p. 24.
159 "Ugovor o financiranju."
160 Nevenka Horvat, "Brodogradnji ipak proračunska dotacija?," *Novi list*, 27 December 1998.
161 Nevenka Horvat, "Brodogradnja i dalje u egzistencijalnoj neizvjesnosti," *Novi list*, 29 May 1999; and "Sanacija i restrukturiranje," 112.
162 L.B.M., "Odobrena pomoć brodogradnji," *Novi list*, 9 November 1999.
163 Horvat, "Brodogradnji ipak proračunska dotacija?"
164 Kersan-Škabić, "Brodogradnja u Europskoj Uniji," 387; and Perić Hadžić and Karačić, "Restrukturiranje hrvatske brodogradnje," 123.
165 Culkin, "Appendix 2," 676–7.
166 Murphy, "China, Philippines," 640.
167 ECORYS, Study on Competitiveness, 32.
168 Culkin, "Appendix 2," 680.
169 Markulinčić and Debeljuh, *Uljanik 1856–2006*, 177–80.
170 Agneza Urošević, "Radolović, Program privatizacije je napravljen i spremni smo za EU," *Poslovni tjednik*, 2 January 2006, https://www.poslovni.hr/hrvatska/radolovic-program-privatizacije-je-napravljen-i-spremni-smo-za-eu-2120.
171 Urošević, "Do danas Uljanik isporučio 250 brodova" (see chapter 1, n. 43).
172 "Koniec z liberalną polityką w gospodarce morskiej," *Głos Wybrzeża*, 28 May 2002, 12.
173 "Strajk w stoczni Gdynia dowodzi, że PRL żyje," *Wprost*, 3 March 2002, 8.
174 "Programowa nacjonalizacja," *Puls biznesu*, 2 February 2004, 8.
175 European Commission, "Commission Decision of 6 November 2008," 6, items 63–5.
176 European Commission, 7, items 272–3.
177 Stott, "Surviving EU Accession."
178 European Commission, "State Aid No C 17/2005," 8.
179 European Commission, "Commission Decision of 6 November 2008," 26, item 232; in addition, the shipyard had liabilities to public providers of more than 423 million złoty.
180 European Commission, "Commission Decision of 6 November 2008," 39, item 299.
181 European Commission, 50, items 383–4.

182 Krzysztof Katka, "Odprawa dla Stoczni Gdynia," *Gazeta Wyborcza*, 30 November 2008.
183 Slowikowska and Graham, "Windmills."
184 OECD Working Party on Shipbuilding, *Imbalances*, 60.
185 Hölscher, Nulsch, and Stephan, "Ten Years after Accession," 314.
186 Kersan-Škabić, "Brodogradnja u Europskoj Uniji," 376.
187 Kersan-Škabić, 385; and Tutić, *EU State Aid Rules*, 24.
188 Bajo and Primorac, "Jesu li brodogradilišta"; and Perić Hadžić and Karačić, "Restrukturiranje hrvatske brodogradnje," 126.
189 Bajo and Primorac, "Jesu li brodogradilišta," 2.
190 Hadžić and Karačić, "Restrukturiranje hrvatske brodogradnje," 128; and Bajo, Primorac, and Hanich, "Financial Performance," 9.
191 See S. Varošanec et al., "Sindikati u panici jer Sanader nije isključio gašenje škverova," *Poslovni dnevnik*, 29 August 2008.
192 Bajo, Primorac, and Hanich, "Financial Performance," 10.
193 "Vlada: Za brodogradilište Uljanik ishoditi rješenje da nije tvrtka u poteškoćama," *Vijesti*, 29 July 2010; and "Kosor: Uljanik će pomoći zatvaranju poglavlja 8," *Poslovni dnevnik*, 3 January 2011; on the financial situation of the shipdyards, see Bajo and Primorac, "Jesu li brodogradilišta," 4.
194 Robert Frank and Milan Pavlović, "Mora se proglasiti stečaj" (interview with Karlo Radolović), *Glas Istre*, 10 March 2019, 7.
195 "Sindikalisti protiv privatizacije Uljanika," *Poslovni dnevnik*, 11 March 2008; Marija Brnić, "Bez ponuda za Staklenike i Uljanik," *Poslovnik dnevnik*, 16 May 2011; and Bajo and Primorac, "Jesu li brodogradilišta," 8.
196 Barukčić, "Uspješno smo završili privatizaciju."
197 European Commission, "Commission Approves Changes to Restructuring Plan."
198 Tholen, "Traditional and New Fields." On the 2021 situation, see "Container Shipping: Perfect Storm," *Economist*, 18 September 2021.
199 See Uljanik's list of deliveries: "ULJANIK Reference List," accessed 16 May 2024, https://uljanik.hr/images/povijest/reference.pdf.
200 Jozo Vrdoljak, "Brodogradnja 2.0," *Nacional*, 3 February 2017, 45–51.
201 Steidel, Daniel, and Yildira, *Shipbuilding Market Developments*, 7.
202 See "Financial Reports."
203 Bajo, Primorac, and Hanich, "Financial Performance," 13.
204 Marina Šunjerga, "Konzultantima plaćali 200.000 € mjesečno," *Večernji list*, 24 August 2018.
205 European Commission, "Commission Clears Rescue Aid."
206 Schumpeter, *Capitalism, Socialism and Democracy*, 82–3 (emphasis in original).
207 "About Us," Crist, accessed 15 March 2024, https://crist.com.pl/about-us,2,en.html.
208 "Elektra Receives Another International Award," FinFerries, archived 23 January 2021 at the Wayback Machine, https://web.archive.org

/web/20210123200439/https://www.finferries.fi/en/news/elektra-receives-another-international-award.html.
209 "Neptun Werft," Wikipedia.de, last updated 8 March 2024, 14:07, https://de.wikipedia.org/wiki/Neptun_Werft.
210 "Experience and Expertise," Neptun Werft, accessed 15 March 2024, https://www.neptunwerft.de/en/company/index.jsp.

3. A Safe Haven?

1 Ther, *Neue Ordnung*.
2 Dunn, *Privatizing Poland*.
3 Hardy, *Poland's New Capitalism*.
4 Ost, *Defeat of Solidarity*.
5 Schumann et al., *Rationalisierung, Krise, Arbeiter*.
6 Stråth, *Politics of De-industrialisation*, 11.
7 "Private Ownership versus Government Ownership," Atlas of European Values, accessed 16 May 2024, https://www.atlasofeuropeanvalues.eu/maptool.html. The most recent data is from 2017.
8 Creswell and Poth, *Qualitative Inquiry and Research Design*.
9 Appel and Orenstein, *From Triumph to Crisis*, 65.
10 Kania, "Przemówienie tow. Stanisława Kani"; and Bartel, *Broken Promises*, 211–12.
11 "Stan i perspektywy wdrażania reformy gospodarczej w Stoczni," APG, 130/464/28, p. 3.
12 "Stan i perspektywy wdrażania reformy gospodarczej w Stoczni," p. 2.
13 Stråth, *Politics of De-industrialisation*.
14 "Uchwała Nr 42/84 Rady Ministrów z dnia 19 marca 1984 r. w sprawie udzielenia pomocy finansowej stoczniom produkcyjnym dla poprawy ich gospodarki oraz zapewnienia realizacji zadań gospodarczych w latach 1984–1985," AAN, 1757/2-136.
15 Kornai, *Socialist System*, 138.
16 Kornai, 152.
17 "Regulamin działania rad nadzorczych," 14 May 1984, AAN, 1757/2-136, p. 2.
18 Horvat, *Yugoslav Economic System*, 234.
19 Brunnbauer, "Globalisierung als Chance."
20 "Zakon o načinu podmirenja obveza Republike u 1979. i 1980. godini preuzetih platformom za rješenje problema brodogradnje i brodarstva," *Narodne Novine* 17/142, 29 April 1978.
21 "Informacija u vezi realizacije obveza općina preuzetih Platformom za rješenje problema brodogradnje i brodarstva u 1982. godini," HDA, f. 280, kut.. 250, p. 2.
22 "Bilješka o vodjenim razgovorima članova delegacije Sabora SRH," 25 May 1979, HDA, f. 280, kut. 174, p. 7.

23 Gligorov, "Socialist Self-Management in Yugoslavia," 17.
24 Tyson, "Liquidity Crises," 289.
25 Ben-Ner and Neuberger, "Planned Market Systems."
26 "Zakon o sanaciji i prestanku organizacija udruženog rada," *Službeni list SFRJ* (1980): 41; see also Schult, *Zwischen Stechuhr und Selbstverwaltung*.
27 Ryšlavy, "Hrvatska brodogradnja danas," 104.
28 R. Radošević, "Uzdah olakšanja, vjera u oživljavanje brodogradnje, ali i poveća doza skepse," *Novi list*, 28 October 1995.
29 Robin Antolović, "Brodogradnja dobila šansu da ponovno bude strateška grana hrvatskog gospodarstva," *Glas Istre*, 4 November 2000, appendix "Tresor," 2.
30 "Wystąpienie pokontrolne," 10 July 1985, ANIK, PCM-41008-9-86, p. 3.
31 Jadwiga Bogdanowicz, "Stocznia Gdynia. Jest 80 mln zł.," *Głos Wybrzeża*, 18 May 2004.
32 Jacek Klein, "Stocznia nie płaci rachunków: Energa zmiejsza dostawy," *Dziennik Bałtycki*, 5 August 2004, 11.
33 Jessop, "Spatial Fixes."
34 Żukowski, "Fabryki-urzędy."
35 "Život i riječ," *Brodograditelj*, 11 March 1976, 1.
36 See Brunnbauer, "Die permanente Transformation."
37 See "Referendum o integraciji."
38 Campbell and Lindberg, "Property Rights," 637.
39 Campbell and Lindberg; and Darko Kojić, "Novi, Jugoslavenski Jadranbrod: Potpisan samoupravni sporazum kojim se objedinujuje cjelokupna jugoslavenska pomorska i riječna brodogradnja," *Brodograditelj*, 27 March 1975.
40 "Prilog raspravi o daljnjem razvoju integracije brodo građevine industrije: Jadranbrod na amandmanskim osnovama," *Brodograditelj*, 7 December 1972, 4–5.
41 Puljar D'Alessio, *Mi gradimo brod, a brod gradi nas*.
42 Sajatović, "Prenosimo," 180.
43 Sajatović, 180.
44 Schulz and Welskopp, "Wieviel kapitalistisches Unternehmen?"
45 Janusz Wikowski, "Wspólnie czy w pojedynkę?," *Głos Stoczniowca*, 4 September 1981, 10.
46 "Plany są realne," *Głos Stoczniowca*, 4 January 1987, 4.
47 Grala, *Reformy gospodarcze w PRL*, 151.
48 "Plany są realne."
49 "Razem czy osobno," *Głos Stoczniowca*, 17 July 1988, 5.
50 Janusz Wikowski, "Nic o nas – bez nas," *Głos Stoczniowca*, 18 February 1990, 7.
51 Doerffer, *Życie i pasje*, 211.
52 Ministerstwo Pracy i Polityki Socjalne, *Pakt o przedsiębiorstwie państwowym*.
53 "Strategija upravljanja i raspolaganja imovinom u vlasništvu Republike Hrvatske za razdoblje od 2013. do 2017.," *Narodne Novine* 94/2121, 18 July 2013.

54 "Uredba o načinu zaštite interesa Republike Hrvatske u postupku pretvorbe društvenog vlasništva u druge oblike vlasništva," *Narodne Novine* 43/811, 24 October 1990, https://narodne-novine.nn.hr/clanci/sluzbeni/1990_10_43_811.html.
55 Gregurek, "Stupanj i učinci privatizacije."
56 D. Kojić, "Brodogradnja – izvozni adut," *Glas Istre*, 19 March 1992, 2.
57 See *OECD Guidelines on Corporate Governance of State-Owned Enterprises*.
58 "Zakon o upravljanju državnom imovinom," *Narodne Novine* 145/3669, 17 December 2010.
59 Grzymala-Busse and Jones Luong, "Reconceptualizing the State"; and Bunce, *Subversive Institutions*.
60 Sznajder Lee, *Transnational Capitalism*.
61 Ekiert, "The State after State Socialism," 291.
62 The term "self-limitation" was chosen following Staniszkis's 1984 analysis of Solidarność's self-limiting revolution; see Staniszkis, *Poland's Self-Limiting Revolution*.
63 Andrzej Kensbok, "Państwowe koło zamachowe," *Rzeczy wspólne* 21/3 (2015): 176.
64 Sabel and Stark, "Planning, Politics, and Shop-Floor Power"; and Kozarzewski, *Polityka właścicielska państwa*.
65 "Ustawa o zasadach wykonywania uprawnień przysługujących Skarbowi Państwa," *Dziennik Ustaw* 106/493, 8 August 1996, 2340–3.
66 Gajdzik, "Steel Industry in Poland."
67 "Narodowy Program Nadzoru Właścicielskiego – Nowy ład korporacyjny w spółkach Skarbu Państwa," 23 June 2010, https://orka.sejm.gov.pl/Biuletyn.nsf/wgskrnr6/SUP-150.
68 "PiS domaga się pomocy od Unii."
69 Szczepański, *Konflikt przemysłowy*.
70 Ost, *Defeat of Solidarity*.
71 Bałtowski, *Gospodarka socjalistyczna w Polsce*.
72 Otto, "Gra o wpływy," 19.
73 Verdery, *What Was Socialism?*; Adam, *Why Did the Socialist System Collapse?*; Lane, *Socialist Industrial State*; Kornai, *Economics of Shortage*; and Kornai, "Resource-Constrained."
74 Sabel and Stark, "Planning, Politics, and Shop-Floor Power," 449.
75 Sabel and Stark, 449.
76 Sabel and Stark, 449.
77 Bieliński and Sadowski, *Ekonomika i organizacja*.
78 For the sake of comparison, the ports and the merchant fleet employed 22,000 people and the fishing industry 33,000.
79 "Eksport przedsiębiorstw przemysłu okrętowego," *Głos Stoczniowca*, 24 January 1988, 6.
80 Sajatović, "Prenosimo."

81 "'Uljanik' deseti izvoznik," *Glas Istre*, 12 March 1985, 3.
82 Tymiński, "Managers in the Command Economy," 165.
83 Kornai, "Soft Budget Constraint," 10.
84 Schierup, *Migration, Socialism*, 150–1.
85 Prinčič, "Direktorska funkcija," 88.
86 Fikfak et al., "To Be a Director," 270.
87 Maciej Borowski, "Lobby na pochylni," *Wybrzeże*, 8 April 1984, 6–7.
88 Ryšlavy, "Hrvatska brodogradnja danas," 105.
89 Stark, "Recombinant Property."
90 Čuvalo, "Hrvatska brodogradnja," 234.
91 Ryšlavy, "Hrvatska brodogradnja danas."
92 Jagoda Vukušić, "Privatizacija: Ključ gospodarskog razvoja," *Glas Istre*, 24 January 1992, 5.
93 Ratko Radošević, "Metlom po brodograđevnom čvoru," *Glas Istre*, 28 February 1994, 8.
94 Brunnbauer, "Globalisierung als Chance."
95 Ratko Radošević, "Država štiti nezasitni profit a ne radnike," *Glas Istre*, 24 November 1995, 20.
96 Vanhuysse, *Divide and Pacify*, 3.
97 Čuvalo, "Hrvatska brodogradnja."
98 D. Grakalić, "SPD i HDZ složni: Samo državna potpora može spasiti hrvatska brodogradilišta," *Glas Istre*, 11 June 2018, 4–5.
99 Dalić, "Budućnost brodogradnje u Hrvatskoj."
100 "Miletić: Odluka Vlade o Uljaniku ispravna i pravo vremena," *Glas Istre*, 22 January 2018, 5.
101 "Predstavnici radnika i sindikata Uljanika krenuli u Zagreb na sastanak s predsjednicom Grabar-Kitarović," *Glas Istre*, 27 March 2019, https://www.glasistre.hr/pula/predstavnici-radnika-i-sindikata-uljanika-krenuli-u-zagreb-na-sastanak-s-predsjednicom-grabar-kitarovic-584305.
102 Quoted in Hodges, "Psychic Landscapes," 69.
103 E.g., Milan Pavlović, "Impresivna platforma 'Apollo' vrijedna 100 milijuna dolara isporučena naručitelju," *Glas Istre*, 10 August 2018, 4.
104 Milan Pavlović, "Vlada mora reći: Uljanik opstaje ili Uljanik ne opstaje," *Glas Istre*, 23 August 2018, 3.
105 Milan Pavlović, "Kermas Energija Danka Končara: S budućim naručiteljima već imamo dogovor za izgradnju brodova u vrijednosti od milijardu dolara," *Glas Istre*, 20 July 2018, 3.
106 Milan Pavlović, "Gianni Rossanda: Predsjednik uprave Uljanika. Brodogradnja nije upitna, ostaje, ali se Uljanik ipak mora mijenjati," *Glas Istre*, 9 February 2018, 5–9.
107 Hodges, "Psychic Landscapes"; and Javeline, *Protest and the Politics of Blame*.

108 OECD Working Party on Shipbuilding, *Imbalances*, 52–5.
109 Ahrens, "Being European."
110 Council of the European Union, "Council Regulation (EC) No 1177/2002."
111 Daniel and Yildiran, "Ship Finance Practices," 44.
112 European Commission, "Communication from the Commission to the Council."
113 Warlouzet, "Collapse of the French Shipyard."
114 Warlouzet.

4. Welded Together

1 Jan Gumiński was a functionary of the All-Polish Trade Union Association (Ogólnopolskie Porozumienie Związków Zawodowych, OPZZ), which was established by the government after Solidarność was banned in 1982. Since 1989, the OPZZ, which leans to the left of the political spectrum, has been the second-largest trade union in Poland.
2 Muehlebach, "Body of Solidarity," 118.
3 Graber Majchrzak, *Arbeit – Produktion – Protest*, 176–95.
4 Dunn, *Privatizing Poland*, 16–17.
5 Mira Čakić, "Evropsko radno vrijeme – za ženu samo gore," *Uljanik*, 27 April 1990, 18.
6 Dunn, *Privatizing Poland*, 171.
7 Negt and Kluge, *Geschichte und Eigensinn*, 3:1154.
8 Polanyi, "Sozialistische Rechnungslegung," 378.
9 Burawoy, "Roots of Domination."
10 A good, almost generic, and well-explored example is the enterprise-based cultural work in the German Democratic Republic: Schuhmann, *Kulturarbeit im sozialistischen Betrieb*.
11 Fonović Cvijanović and Vitković Marčeta, "Language of the Socialist Man."
12 Inowrocław is a small town some 230 kilometres to the south of Gdynia.
13 Kornai, *Economics of Shortage*. For an insightful application of this concept to the Polish context, see Mazurek, *Społeczeństwo kolejki*.
14 The concept of the "social vaccum" was coined in 1979 by the eminent Polish sociologist Stefan Nowak in his influential article "System wartości społeczeństwa polskiego." For a critical debate of the (mis)use of this concept, see Pawlak, "Sociological Vacuum."
15 These are quotes from casual conversations during participant observation, which is why we do not mention the initials of the respondents.
16 Hann, *Property Relations*.

17 See Mitchell, "Work Authority in Industry"; and Sabel and Stark, "Planning, Politics, and Shop-Floor Power."
18 Letter to SIV, Generalni sekretar, broj 03-291-1979, 9 January 1979, HDA, f. 280 (IVS), kut. 161, br. 7.
19 "Izdvajane iz brutto osobnog dohotka," *Mali informator*, 8 April 1976, 1.
20 See, e.g., "Upravljanje sredstvima samodoprinosa," *Mali informator*, 14 June 1979, 2.
21 "Plan zajedničke potrošnje," *Mali informator*, 20 February 1989, 2.
22 Marija Rosanda, "Stambena Problematika BI Uljanika," [1986], Uljanik Company Archive, Tvornički odbor sindikata, Pošta, no signature, p. 1.
23 "Popis stanara SP 'Uljanik,'" Pula, 31 March 1992, HFP Archive, no signature.
24 "Uljanik ne traži dominaciju u gradu a svoja prava," *Mali informator*, 27 March 1989, 1–3.
25 "Rocznik Statystyczny 1981," 24 April 1982, APG, 130/2365.
26 "Przedsiębiorstwo hotelowo-turystyczne S. A.," *Cała Naprzód*, nos. 7–8 (1993): 5.
27 Szymon Szadurski, "Odprawienie z kwitkiem," *Dziennik Bałtycki*, 26 August 2003, 5.
28 Szymon Szadurski, "Przychodnia nie dla stoczniowców," *Dziennik Bałtycki*, 24 June 2004.
29 S.W., "Dopełnił się akt likwidacji. Zakończenie roku w ZST z udziałem poseł M. Zwiercan i Sztandaru Solidarności," *Młoda Gdynia*, 27 June 2016, https://dziendobrypomorze.pl/pl/11_wiadomosci/15672_dopelnil-sie-akt-likwidacji-zakonczenie-roku-w-zst-z-udzialem-posel-m-zwiercan-i-sztandaru-solidarnosci.html.
30 Crowley and Ost, *Workers after Workers' States*.
31 Reljanović, "Normative Position of Trade Unions."
32 Musić, *Yugoslav Working Class*.
33 Social anthropologist Ognjen Kojanic encountered very similar narratives in his research on railway workers in Zaječar, Serbia: Kojanic, "Nostalgia as a Practice."
34 "Jednako se plaća rad i nerad," *Uljanik*, 8 February 1989, 10.
35 "Jednako se plaća rad i nerad," 15.
36 Burawoy, "Roots of Domination," 199; cf. Dunn, *Privatizing Poland*, 169.
37 Hodges, "Between Emic and Etic." In their vernacular, our respondents spoke of "pravi kapitalizam je pravedan" and "sustavno uništavanje industrije."
38 Von Hirschhausen and Bitzer, *Globalization of Industry and Innovation*.
39 The fordist organization of labour in capitalism was not much different; see Muehlebach, "Body of Solidarity," 98.
40 Kalb, "Conversations with a Polish Populist," 215.
41 See also Dunn, *Privatizing Poland*, 160, which comes to similar conclusions. Verdery argues that the socialist state presented itself as a "parent-state": Verdery, *What Was Socialism?*, 61–82.

42 Inspired by the work of Edward Banfield, Polish sociologists identified "familialism" as a crucial characteristic of Polish society during late socialism, with all its negative effects on public life (described also by Banfield on the example of Southern Italy). See, e.g., Tarkowska and Tarkowski, "'Amoralny familizm.'"
43 Fidelis, *Women, Communism, and Industrialization*.
44 Graber Majchrzak, "Gdańsk Shipyard," 387.
45 See Hrobat Virloget, Gousseff, and Corni, *At Home but Foreigners*.
46 Burawoy, "Roots of Domination," 199.
47 Vodopivec, "Past for the Present," 228.
48 Goffman, *Asylums*.
49 Foucault, "Of Other Spaces."
50 E.g., "Snježana Milivojević: Država kao hobotnica," *Danas*, 2 February 2017, https://www.danas.rs/vesti/drustvo/snjezana-milivojevic-drzava-kao-hobotnica/.
51 Markulinčić and Debeljuh, *Uljanik 1856–2006*.
52 Vodopivec, "Past for the Present," 214.
53 Dunn, "Slick Salesmen," 125.
54 Gardawski, *Polacy pracujący*; and Marody, *Jednostka po nowoczesności*.
55 See Raphael, *Jenseits von Kohle und Stahl*.
56 Dunn, *Privatizing Poland*, 169.
57 See Dolan and Rajak, *Corporate Social Responsibility*.
58 Brunnbauer, "Die permanente Transformation."

5. Added Value

1 See Schulz and Welskopp, "Wieviel kapitalistisches Unternehmen?"
2 Santos, "From Iron to the Industrial Cloud," 97.
3 Kideckel, *Getting By in Postsocialist Romania*.
4 Raphael, *Jenseits von Kohle und Stahl*.
5 Scribner, *Requiem for Communism*, 6.
6 Todorova, *Remembering Communism*.
7 A useful summary of the relationship between the workers and the state is provided in Pittaway, *Eastern Europe*; and Leszczyński, *Ludowa historia Polski*.
8 Scribner, *Requiem for Communism*, 20 (emphasis in original).
9 Verdery, *What Was Socialism?*, 24.
10 Pittaway, *Eastern Europe*.
11 See, e.g., Todorova, *Remembering Communism*; Hann, *Postsocialism*; Morris, *Everyday Post-Socialism*; and Leyk and Wawrzyniak, *Cięcia*.
12 Bonfiglioli, *Women and Industry*; and Morris, *Everyday Post-Socialism*.
13 Kalb, "Introduction," 25.
14 Muehlebach, "Body of Solidarity," 101.
15 See, e.g., Kofti, "Moral Economy of Flexible Production."

16 Petrović, "Socialist Workers in Serbia," 128.
17 Marković, "Der Sozialismus."
18 Crowley and Ost, "Introduction," 3–4.
19 Scribner, *Requiem for Communism*, 3.
20 Kideckel, *Getting By in Postsocialist Romania*, 114.
21 Dunn, "Slick Salesmen," 133–4.
22 Quoted in Kalb, "Conversations with a Polish Populist," 213.
23 Ost, "Workers and the Radical Right," 118.
24 Crowley and Ost, *Workers after Workers' States*, 3–4.
25 Ost, *Defeat of Solidarity*.
26 Scribner, *Requiem for Communism*, 16.
27 Ost, *Defeat of Solidarity*, 88–9.
28 Ost, 83–4.
29 Butković, Czarzasty, and Mrozowicki, "Gains and Pitfalls of Coalitions," 47.
30 Kokanović, "Cost of Nationalism," 145; and Butković, Czarzasty, and Mrozowicki, "Gains and Pitfalls of Coalitions," 47–8.
31 Kokanović, "Cost of Nationalism," 147.
32 Grdešić, "Tranzicija, sindikati i političke elite," 123.
33 Kokanović, "Cost of Nationalism," 142–3; cf. Kokanović, "Croatian Labour Realities."
34 Peračković, "Promjene u strukturi zanimanja."
35 Mihaljević, "Industrial Meltdown in Croatia."
36 Morris, *Everyday Post-Socialism*.
37 "Za pohvalu," *Uljanik*, 27 April 1990, 30.
38 K[atica] Š[ipura], "Treba voljeti posao," *Uljanik*, 27 April 1990, 27.
39 Uljanik, *Informativni Priručnik*.
40 Schuster, *Workforce Divided*; and Varela, Murphy, and van der Linden, *Shipbuilding and Ship Repair Workers*.
41 Petrović, "Socialist Workers in Serbia," 142.
42 Muehlebach, "Body of Solidarity," 123.
43 Dunn, *Privatizing Poland*.
44 Burawoy, *Politics of Production*.
45 See, e.g., "Samoupravni sporazum o zajedničkim osnovama i mjerilima za uređivanje prava, obaveza i odgovornosti radnika u radnom odnosu," in *Dodatak Vjesniku Uljanika*, 7 December 1978, esp. §§ 52–72.
46 Brunnbauer and Nonai, "Finding Workers."
47 Petrungaro, "Ethics of Work."
48 Karlo Radolović, "Radimo više, kako bi nam bilo bolje," *Uljanik*, 30 December 1980, 2.
49 Quoted in Vladimir Filipović, "Dobro poslovanje uljepšava obljetnicu," *Uljanik*, 1 January 1986, 28.
50 Matošević, "Tehnička događajnica," 196.
51 See Vodopivec, *Berge Istra*.

52 For evidence of this collective pride in the public launching of new ships, see "'Sennar': novi uspjeh brodogradilišta," *Uljanik*, no. 10, October 1961, 2–4; and "Svečano porinut putnički brod za Brazil 'Anna Nery,'" *Uljanik*, nos. 11–12, November–December 1961, 2.
53 Stråth, *Politics of De-industrialisation*, 21.
54 For similarities with the self-perception of workers in two other Yugoslav factories with a strong global orientation, see Spaskovska and Calori, "Nonaligned Business World."
55 This is evident from articles that appeared in the industry magazine *Brodogradnja*, especially those that reproduced foreign reports, such as Lloyd's.
56 Vera Cukon, "Kakve perspektive otvara ženama petogodišnji plan," *Uljanik*, nos. 2–3, February–March 1961, 18; and S. Maršić, "Još uvijek premalo dječjih vrtića i jaslica," *Uljanik*, 1 June 1981, 18–19.
57 "Žena kao radnica u BI Uljaniku," *Uljanik*, 3 March 1986, 15–17; cf. Bonfiglioli, *Women and Industry*; and Vodopivec, "Past for the Present."
58 Muehlebach, "Body of Solidarity," 98.
59 Stanić, "'Jedan od najtežih dana u Uljaniku!'"
60 Vodopivec, "Past for the Present," 221.
61 Stocznia Gdynia in particular achieved tragic notoriety when, in December 1970, the army fired on workers during strikes in Gdańsk and Gdynia, resulting in ten fatalities. This is a symbolic event that has been commemorated many times in Polish literature and films.
62 "Siguran posao i plaće – to je ono što nas bogati," *Mali informator*, 10 December 1999, 1–2.
63 Matošević, "Tehnička događajnica," 209.
64 See Fikfak, Prinčič, and Turk, *Biti direktor v času socializma*.
65 For examples of the published opinions of a skilled and an unskilled worker, see Albino Padien, "Dosta nepravilnosti," and Zora Buić, "Treba samo više raditi," *Uljanik*, 8 March 1990, 17–18. An even more radical position was voiced by the new vice director of the Brodogradilište (Shipbuilding) department, Jakov Tomičić: "Slabe osnove za optimizam," *Uljanik*, 27 April 1990, 20–1.
66 Discussions on this topic had already begun in the 1980s; see, e.g., Franko Kopal, "Organizacija i kadrovi u brodogradnji," *Uljanik*, 8 February 1982, 14–15.
67 See Sučec, "Tko je Gianni Rossanda."
68 Neither in the European Solidarność Centre (Europejskie Centrum Solidarności) nor in the newly established Institute for the Heritage of Solidarność (Instytut Dziedzictwa Solidarności) does shipbuilding take centre stage. Instead, the narrative in both institutions, despite their different political leanings, highlights the fight against communist rule.
69 Morris, *Everyday Post-Socialism*.

70 Cowie and Heathcott, *Beyond the Ruins*.
71 Morris, *Everyday Post-Socialism*, 241.
72 The Pula Travel Guide provides an apt illustration: "Lightning Giants Pula," Pula-Croatia, 6 June 2015, https://www.pulacroatia.net/pula-croatia/lightning-giants-pula/.
73 Diagnoza Społeczna (Social Diagnosis), the most reliable public opinion barometer in Poland, repeatedly ranked Gdynia as the city with the highest share of respondents expressing satisfaction with their place of residence (more than 80 per cent). See Czapiński and Panek, *Social Diagnosis 2015*.

6. Keel Up?

1 Orjana Antešić, "Uskoro isplata 502 milijuna od prodaje jaružala Jan De Nulu, Vlada na ovom poslu gubi 400 milijuna," *Novi List*, 23 May 2020.
2 Bakša, "Wedding of Willem van Rubroeck."
3 See "History," Remontowa Shiprepair Yard, accessed 28 March 2024, https://www.remontowa.com.pl/about/history/.
4 World Bank Open Data, accessed 28 March 2024, https://data.worldbank.org.
5 Gourdon, "Market-Distorting Factors," 16.
6 Steidel, Daniel, and Yildira, *Shipbuilding Market Developments*, 17.
7 Kalouptsidi, "Detection and Impact"; and Hossain, Zakaria, and Sarkar, "SWOT Analysis."
8 Gourdon, "Market-Distorting Factors," 25.
9 Hann, *Repatriating Polanyi*.
10 Hann, *Repatriating Polanyi*.
11 Warlouzet, "Collapse of the French Shipyard."
12 In this respect, our book complements Tooze, *Crashed*.
13 The importance of government intervention for the outcome of processes of deindustrialization is exemplified by the Great Lakes region of North America, where Canada was much more successful than the United States at mitigating the social costs of this structural change through social dialogue and by preventing industrial enterprises from disappearing completely; see Steven C. High's excellent study *Industrial Sunset*.
14 Krastev and Holmes, *Light That Failed*.
15 Kalb, "Conversations with a Polish Populist," 211.
16 Muehlebach, "Body of Solidarity."
17 Boym, *Future of Nostalgia*; see also her definition of the concept in "Nostalgia."
18 On Germany, see Schumann et al., *Rationalisierung, Krise, Arbeiter*; Alheit and Dausien, *Arbeiterbiographien*; and Grüner and Mecking, *Wirtschaftsräume und Lebenschancen*.
19 Morris, *Everyday Post-Socialism*, 240.

20 Kofti, "'Communists' on the Shop Floor."
21 On these temporalities, see Doering-Manteuffel, Raphael, and Schlemmer, *Vorgeschichte der Gegenwart*.
22 This was recognized as early as in 1991 in Offe, "Das Dilemma der Gleichzeitigkeit," 283.
23 Bartel, *Broken Promises*.
24 On the dynamnics between political action, unintended effects, and political responses to them, see Creed, *Domesticating Revolution*.
25 Campbell and Hall, *Paradox of Vulnerability*.
26 Pula, *Globalization*.
27 Unkovski-Korica, *Economic Struggle for Power*.
28 Sajatović, "Prenosimo," 178.
29 Appel and Orenstein, *From Triumph to Crisis*, 65, 180.
30 Brunnbauer and Hodges, "Long Hand of Workers' Ownership."
31 Tooze, *Crashed*.
32 Ost, "Workers and the Radical Right," 115.
33 Ost; and Kalb, "Conversations with a Polish Populist."
34 Ost, "Workers and the Radical Right," 121.
35 "Results of voting in 2023 elections for Sejm," National Electoral Commission, accessed 11 April 2024, https://sejmsenat2023.pkw.gov.pl/sejmsenat2023/en/sejm/wynik/pow/226200.
36 "Results of voting in 2023 elections for Senate City county Gdynia," accessed 11 April 2024, https://sejmsenat2023.pkw.gov.pl/sejmsenat2023/en/senat/wynik/pow/226200.
37 Official election results at "Izbori za zastupnike u hrvatski sabor," Arhiva izbora Republike Hrvatske, 5 July 2020, https://www.izbori.hr/arhiva-izbora/index.html#/app/parlament-2020.
38 Witold M. Orłowski, "Smutna historia naszych stoczni," *Gazeta Wyborcza*, 18 July 2008.
39 Appel and Orenstein, *From Triumph to Crisis*, 24.
40 Appel and Orenstein, 65, 180.
41 Sznajder Lee, *Transnational Capitalism*; and Trappmann, *Fallen Heroes*.
42 Verdery, *What Was Socialism?*
43 Milan Pavlović, "Vratio se dredžer," *Glas Istre*, 6 August 2020.

Postscript

1 "Die maritime Branche sucht den Weg aus der Krise," *Süddeutsche Zeitung*, 10 May 2021.
2 SEA Europe, "Open Letter."
3 SEA Europe, "Open Letter."
4 "Container Shipping: Perfect Storm," *Economist*, 18 September 2021, 61–2.

Bibliography

Archival Sources

Archive of the former Hrvatski fond za privatizaciju (Croatian Privatization Fund, Zagreb; HFP Archive)

[no inventory and signatures]

Archiwum Akt Nowych (Archive of New Documents, Warsaw; AAN)

1757: Ministerstwo Hutnictwa i Przemysłu Maszynowego (Ministry of the Steel and Machine-Building Industry)
1786: Ministerstwo Przemysłu (Ministry of Industry)

Archiwum Najwyższej Izby Kontroli (Archive of the Supreme Audit Office, Warsaw; ANIK)

[no inventory and signatures]

Hrvatski državni arhiv (Croatian State Archive, Zagreb; HDA)

280: Izvršno vijeće sabora (Sabor Executive Council, i.e., government of the socialist republic)
1286: Savez Sindikata Jugoslavije. Vijeće Saveza sindikata Hrvatske (Union of the Trade Unions of Yugoslavia, Council of the Trade Unions of Croatia)
1321: Sindikat metalskih radnika Jugoslavije. Republički odbor za Hrvatsku (Syndicate of Metalworkers of Yugoslavia, Republican Committee for Croatia)
1398: Sindikat Metalaca Hrvatske (Syndicate of Metalworkers of Croatia)

1596: Republički sekretarijat za privredu Socijalističke Republike Hrvatske (Republican Economic Secretariate [i.e., ministry] of the Socialist Republic of Croatia)
1691: Samoupravna interesna zajednica Hrvatske za ekonomske odnose s inozemstvom (Self-Managed Interest Community of Croatia for Economic Relations with Foreign Countries)

Archiwum Państwowe w Gdańsku (Gdańsk State Archive, APG)

130: Stocznia Gdynia (Gdynia Shipyard)
1091: Zjednoczenie Przemysłu Okrętowego (Association of the Shipbuilding Industry)
2371: Zrzeszenie Przedsiębiorstw Przemysłu Okrętowego (Association of Shipbuilding Businesses)

State Archive Pazin, Pazin (DPA)

390: Gradski komitet Saveza komunista Hrvatske Pula (Pula City Commitee of the League of Communists of Croatia)
394: Općinska konferencija Saveza komunista Hrvatske Pula (Pula Municipal Commitee of the League of Communists of Croatia)

Uljanik Company Archive, Pula

Radnički savjeti (workers' councils)
Zapisnici Poslovnog savjeta (minutes of the Executive Council)
Tvornički odbor sindikata (trade union)

Newspapers

Borba
Brodograditelj
Cała Naprzód
Danas
Dziennik Bałtycki
Dziennik Ustaw
Economist
Gazeta Wyborcza
Glas Istre
Głos Stoczniowca
Głos Wybrzeża
Mali Informator

Młoda Gdynia
Nacional
Narodne Novine
Nedjeljna Dalmacija
Novi list
Peking Review
Politika
Polityka
Poslovni dnevnik
Poslovni tjednik
Privjredni Vjesnik
Puls biznesu
Rzeczy wspólne
Start
Süddeutsche Zeitung
Uljanik
Večernje novosti
Vjesnik
Vjesnik u srijedu (VUS)
Wprost
Wybrzeże

Secondary Literature

Abrams, Lynn. *Oral History Theory*. 2nd ed. London: Routledge, 2016. https://doi.org/10.4324/9781315640761.
Adam, Jan. *Why Did the Socialist System Collapse in Central and Eastern European Countries? The Case of Poland, the Former Czechoslovakia and Hungary*. Basingstoke: Macmillan, 1996. https://doi.org/10.1007/978-1-349-24239-9.
Ahrens, Ralf. "The Importance of Being European: Airbus and West German Industrial Policy from the 1960s to the 1980s." *Journal of Modern European History* 18, no. 1 (February 2020): 63–78. https://doi.org/10.1177/1611894419894475.
Alheit, Peter, and Bettina Dausien. *Arbeiterbiographien. Zur thematischen Relevanz der Arbeit in proletarischen Lebensläufen. Eine exemplarische Untersuchung im Rahmen der "biographischen Methode."* 2nd ed. Bremen: University of Bremen, 1985.
Alheit, Peter, and Hanna Haack. *Die vergessene "Autonomie" der Arbeiter. Eine Studie zum frühen Scheitern der DDR am Beispiel der Neptunwerft*. Berlin: Karl Dietz, 2004.
Appel, Hilary, and Mitchell A. Orenstein. *From Triumph to Crisis: Neoliberal Economic Reform in Postcommunist Countries*. Cambridge: Cambridge University Press: 2018. https://doi.org/10.1017/9781108381413.

Babić, Mate, and Emil Primorac. "Some Causes of the Growth of the Yugoslav External Debt." *Soviet Studies* 38, no. 1 (1986): 69–88. https://doi.org/10.1080/09668138608411623.

Bajo, Anto, and Marko Primorac. "Jesu li brodogradilišta prepreka fiskalnoj konsolidaciji u Hrvatskoj?" Newsletter 64, Institut za javne financije, 2011. https://www.ijf.hr/upload/files/file/newsletter/64.pdf.

Bajo, Anto, Marko Primorac, and Martin Hanich. "The Financial Performance, Restructuring and Privatisation of the Shipyards in the Republic of Croatia." CIRIEC Working Papers 2018/02, CIRIEC International, Université de Liège, 2018.

Bakša, Denis. "Wedding of Willem van Rubroeck and Adriatic Sea." Istriago.net, accessed 28 March 2024. https://www.istriago.net/de/wedding-of-willem-van-rubroeck-and-adriatic-sea-in-pula/.

Balcerowicz, Leszek. *800 dni. szok kontrolowany.* Compiled by Jerzy Baczyński. Warsaw: BGW, 1992.

Bałtowski, Maciej. *Gospodarka socjalistyczna w Polsce. Geneza – rozwój – upadek.* Warsaw: PWN, 2009.

Barrett, Elizabeth. "The Role of Informal Networks in the Privatisation Process in Croatia." In *Restructuring of the Economic Elites after State Socialism: Recruitment, Institutions and Attitudes*, edited by David Lane, György Lengyel, and Jochen Tholen, 211–40. Stuttgart: ibidem, 2007.

Bartel, Fritz. *The Triumph of Broken Promises: The End of the Cold War and the Rise of Neoliberalism.* Cambridge, MA: Harvard University Press, 2022. https://doi.org/10.4159/9780674275805.

Barukčić, Marina. "Uspješno smo završili privatizaciju brodogradilišta Uljanik." tportal, 26 July 2012. https://www.tportal.hr/vijesti/clanak/uspjesno-smo-zavrsili-privatizaciju-brodogradilista-uljanik-20120726.

Baučić, Ante. "Početak restrukturiranja brodogradilišta zakonom o sanaciji." *Brodogradnja* 43, no. 3 (1995): 201–2.

Ben-Ner, Avner, and Egon Neuberger. "The Feasibility of Planned Market Systems: The Yugoslav Visible Hand and Negotiated Planning." *Journal of Comparative Economics* 14, no. 4 (December 1990): 768–90. https://doi.org/10.1016/0147-5967(90)90052-B.

Berend, Iván T. *From the Soviet Bloc to the European Union: The Economic and Social Transformation of Central and Eastern Europe since 1973.* Cambridge: Cambridge University Press, 2009. https://doi.org/10.1017/CBO9780511806995.

Bieliński, Jerzy, and Zdzisław Sadowski. *Ekonomika i organizacja przemysłu okrętowego.* Gdańsk: Uniwersytet Gdański, 1985.

"Big Mac nachodzi." *Polska Kronika Filmowa*, 17 June 1992. YouTube video, 2:34. https://www.youtube.com/watch?v=0wnA6QxuGEw.

Bitzer, Jürgen, and Christian von Hirschhausen. "The Shipbuilding Industry in the East and West: Industry Dynamics, Science and Technology Policies and Emerging Patterns of Cooperation." DIW Discussion Papers 151, Deutsches Institut für Wirtschaftsforschung, Berlin, 1997. https://hdl.handle.net/10419/61530.

Bohle, Dorothee, and Béla Greskovits. *Capitalist Diversity on Europe's Periphery*. Ithaca, NY: Cornell University Press, 2012.

Bonfiglioli, Chiara. *Women and Industry in the Balkans: The Rise and Fall of the Yugoslav Textile Sector*. London: I.B. Tauris, 2021. https://doi.org/10.5040/9781838600778.

Bösch, Frank. *Mediengeschichte. Vom asiatischen Buchdruck zum Fernsehen*. Frankfurt am Main: Campus, 2011.

Boym, Svetlana. *The Future of Nostalgia*. New York: Basic Books, 2001.

– "Nostalgia." Atlas of Transformation, accessed 28 March 2024. http://monumenttotransformation.org/atlas-of-transformation/html/n/nostalgia/nostalgia-svetlana-boym.html.

Brunnbauer, Ulf. "Die permanente Transformation. Vom Nutzen des Durchwurstelns und seinen Grenzen am Beispiel der Werft Uljanik in Pula seit den 1970er Jahren." In *Transformation als soziale Praxis. Mitteleuropa nach dem Boom*, edited by Dierk Hoffmann and Ulf Brunnbauer, 21–38. Berlin: Metropol, 2020.

– "Globalisierung als Chance. Die vielen Leben der Schiffswerft Uljanik in Pula." In *Erfahrungs- und Handlungsräume. Gesellschaftlicher Wandel in Südosteuropa seit dem 19. Jahrhundert zwischen dem Lokalen und dem Globalen. Festschrift für Wolfgang Höpken*, edited by Heike Karge, Ulf Brunnbauer, Claudia Weber, 95–117. Munich: De Gruyter Oldenbourg, 2017.

Brunnbauer, Ulf, and Andrew Hodges. "The Long Hand of Workers' Ownership: Performing Transformation in the Uljanik Shipyard in Yugoslavia/Croatia, 1970–2018." *International Journal of Maritime History* 31, no. 4 (November 2019): 860–78. https://doi.org/10.1177/0843871419874003.

Brunnbauer, Ulf, and Visar Nonai. "Finding Workers to Build Socialism: Recruiting for Steel Factories in Bulgaria and Albania." In *Labor in State-Socialist Europe, 1945–1989: Contributions to a History of Work*, edited by Marsha Siefert, 73–98. Budapest: Central European University Press, 2020. https://doi.org/10.1515/9789633863381-007.

Bruno, Lars, and Stig Tenold. "The Basis for South Korea's Ascent in the Shipbuilding Industry, 1970–1990." *Mariner's Mirror* 97, no. 3 (2011): 201–17. https://doi.org/10.1080/00253359.2011.10708948.

Bunce, Valerie. *Subversive Institutions: The Design and the Destruction of Socialism and the State*. Cambridge: Cambridge University Press, 1999. https://doi.org/10.1017/CBO9780511816178.

Burawoy, Michael. *The Politics of Production*. London: Verso Books, 1985.

– "The Roots of Domination: Beyond Bourdieu and Gramsci." *Sociology* 46, no. 2 (April 2012): 187–206. https://doi.org/10.1177/0038038511422725.

Butković, Hrvoje, Jan Czarzasty, and Adam Mrozowicki. "Gains and Pitfalls of Coalitions: Societal Resources as Sources of Trade Union Power in Croatia and Poland." *European Journal of Industrial Relations* 29, no. 1 (March 2023): 43–61. https://doi.org/10.1177/09596801221138776.
Bykova, Alexandra, Beate Muck, Renate Prasch, Monika Schwarzhappel, and Galina Vasaros. *wiiw Handbook of Statistics 2012: Central, East and Southeast Europe*. Vienna: wiiw, 2012.
Calic, Marie-Janine. "The Beginning of the End: The 1970s as a Historical Turning Point in Yugoslavia." In *The Crisis of Socialist Modernity: The Soviet Union and Yugoslavia in the 1970s*, edited by Marie-Janine Calic, Dietmar Neutatz, and Julia Obertreis, 66–86. Göttingen: Vandenhoeck & Ruprecht, 2011.
Campbell, John L., and John A. Hall. *The Paradox of Vulnerability: States, Nationalism, and the Financial Crisis*. Princeton, NJ: Princeton University Press, 2017. https://doi.org/10.23943/princeton/9780691163260.001.0001.
Campbell, John L. and Leon N. Lindberg. "Property rights and the organization of economic activity by the state." *American Sociological Review* 55, no. 5 (October 1990): 634–47. https://doi.org/10.2307/2095861
CIA Directorate of Intelligence. *Yugoslavia: Key Questions and Answers on the Debt Crisis. An Intelligence Assessment*. January 1984.
Collins, Gabriel, and Michael C. Grubb. *A Comprehensive Survey of China's Dynamic Shipbuilding Industry*. Newport, RI: US Naval War College, 2008.
Colton, Tim, and LaVar Huntzinger. *A Brief History of Shipbuilding in Recent Times*. Alexandria, VA: Center for Naval Analyses, 2002.
Council of the European Union. "Council Regulation (EC) No 1177/2002 of 27 June 2002 Concerning a Temporary Defensive Mechanism to Shipbuilding." 27 June 2002. https://eur-lex.europa.eu/legal-content/EN/TXT/?uri=CELEX%3A32002R1177.
Cowie, Jefferson, and Joseph Heathcott, eds. *Beyond the Ruins: The Meanings of Deindustrialization*. Ithaca, NY: ILR Press, 2003.
Creed, Gerald W. *Domesticating Revolution: From Socialist Reform to Ambivalent Transition in a Bulgarian Village*. University Park: Pennsylvania State University Press, 1997.
Creswell, John W., and Cheryl N. Poth. *Qualitative Inquiry and Research Design: Choosing among Five Approaches*. 4th ed. London: SAGE, 2017.
Crowley, Stephen, and David Ost. "Introduction." In Crowley and Ost, *Workers after Workers' States*, 1–12.
Crowley, Stephen, and David Ost, eds. *Workers after Workers' States: Labor and Politics in Postcommunist Eastern Europe*. Lanham, MD: Rowman & Littlefield, 2001.
Culkin, Victoria. "Appendix 2. Shipbuilding in 2013: An Analysis of Shipbuilding Statistics." In Varela, Murphy, and van der Linden,

Shipbuilding and Ship Repair Workers, 675–82. https://doi.org/10.1515
/9789048530724-028.
Čuvalo, Milan. "Hrvatska brodogradnja. Stanje i perspektive." *Brodogradnja*
41, no. 4 (1993): 231–4.
– "Program restrukturiranja hrvatske brodograđevne industrije." *Brodogradnja*
43, no. 3 (1995): 204–14.
– "Projekt – rekonstrukcija sustava hrvatske brodogradnje." *Brodogradnja* 43,
no. 1 (1995): 13–14.
Czapiński, Janusz, and Tomasz Panek, eds. *Social Diagnosis 2015: Objective and
Subjective Quality of Life in Poland*. Warsaw: Rada Monitoringu Społecznego,
2015. http://www.diagnoza.com/pliki/raporty/Diagnoza_raport_2015.pdf.
Dale, Gareth. *Karl Polanyi: A Life on the Left*. New York: Columbia University
Press, 2016. https://doi.org/10.7312/dale17608.
– *Karl Polanyi: The Limits of the Market*. Malden, MA: Polity Press, 2010.
Dalić, Martina. "Budućnost brodogradnje u Hrvatskoj." Presentation at the
conference "The Future of Shipbuilding in Croatia," Zagreb, 27 March 2018.
Daniel, Laurent, and Cenk Yildiran. "Ship Finance Practices in Major
Shipbuilding Economies." OECD Science, Technology and Industry Policy
Papers 75, 2019. https://www.oecd-ilibrary.org/science-and-technology
/ship-finance-practices-in-major-shipbuilding-economies_e0448fd0-en.
Djeković, Liljana. "Außenwirtschaftssystem und Außenwirtschaftsreformen
in Jugoslawien." In *Außenwirtschaftssysteme und Außenwirtschaftsreformen
sozialistischer Länder. Ein intrasystemarer Vergleich*, edited by Maria
Haendcke-Hoppe, 65–86. Berlin: Duncker und Humblot, 1988.
Doerffer, Jerzy Wojciech. *Życie i pasje. Wspomnienia*. Vol. 4, *Emerytura*. Gdańsk:
Fundacja Promocji Przemysłu Okrętowego i Gospodarki Morskiej, 2008.
Doering-Manteuffel, Anselm, and Lutz Raphael, eds. *Nach dem Boom.
Perspektiven auf die Zeitgeschichte seit 1970*. Göttingen: Vandenhoeck &
Ruprecht, 2008.
Doering-Manteuffel, Anselm, Lutz Raphael, and Thomas Schlemmer, eds.
Vorgeschichte der Gegenwart. Dimensionen des Strukturbruchs nach dem Boom.
Göttingen: Vandenhoeck & Ruprecht, 2016. https://doi.org/10.13109
/9783666300783.
Dolan, Catherine, and Dinah Rajak, eds. *The Anthropology of Corporate Social
Responsibility*. Oxford: Berghahn Books, 2016. https://doi.org/10.3167
/9781785330711.
Dunn, Elizabeth. *Privatizing Poland: Baby Food, Big Business, and the Remaking of
Labor*. Ithaca, NY: Cornell University Press, 2004.
– "Slick Salesmen and Simple People: Negotiated Capitalism in a Privatized
Polish Firm." In *Uncertain Transition: Ehnographies of Change in the
Postsocialist World*, edited by Christopher M. Hann, 125–50. Lanham, MD:
Rowman & Littlefield, 1999.

ECORYS. *Study on Competitiveness of the European Shipbuilding Industry: Final Report.* Rotterdam: ECORYS, 2009. https://ec.europa.eu/docsroom/documents/10506/attachments/1/translations/en/renditions/pdf.

Ekiert, Grzegorz. "The State after State Socialism: Poland in Comparative Perspective." In *The Nation-State in Question*, edited by Thazha V. Paul, Gilford J. Ikenberry, and John A. Hall, 291–320. Princeton, NJ: Princeton University Press, 2003. https://doi.org/10.2307/j.ctv173f30r.16.

European Commission. "Commission Decision of 6 November 2008 on State Aid C 17/05 (ex N 194/05 and PL 34/04) Granted by Poland to Stocznia Gdynia." 2010 O.J. (L 33) 1. https://eur-lex.europa.eu/legal-content/en/ALL/?uri=CELEX:32010D0047&qid=1607037426810.

— "Communication from the Commission to the Council, the European Parliament, the European Economic and Social Committee and the Committee of the Regions – LeaderSHIP 2015 – Defining the Future of the European Shipbuilding and Repair Industry – Competitiveness through Excellence." 21 November 2003. https://eur-lex.europa.eu/legal-content/GA/TXT/?uri=CELEX:52003DC0717.

— "State Aid: Commission Approves Changes to Restructuring Plan of Croatian Shipyard 3. Maj." News release, 19 June 2013. http://europa.eu/rapid/press-release_IP-13-565_en.htm.

— "State Aid: Commission Clears Rescue Aid for Croatian Shipbuilder Uljanik." News release, 22 January 2018. http://europa.eu/rapid/press-release_IP-18-391_en.htm.

— "State Aid – Poland – State Aid No C 17/2005 (ex PL 34/2004 and N 194/2005) and C 18/2005 (ex N 438/04) – Restructuring Aid to Gdynia Shipyard; Restructuring Aid to Gdansk Shipyard." 2005 O.J. (C 220) 7. https://eur-lex.europa.eu/legal-content/EN/TXT/?uri=CELEX:52005XC0908(03).

Ferguson, Niall, Charles S. Maier, Erez Manela, and Daniel J. Sargent, eds. *The Shock of the Global: The 1970s in Perspective.* Cambridge, MA: Harvard University Press, 2010. https://doi.org/10.2307/j.ctvrs8zfp.

Fidelis, Malgorzata. *Women, Communism, and Industrialization in Postwar Poland.* Cambridge: Cambridge University Press, 2010.

Fikfak, Jurij, Jože Prinčič, and Jeffrey D. Turk, eds. *Biti direktor v času socializma. Med idejami in praksami.* Ljubljana: Založba ZRC SAZU, 2008. https://doi.org/10.3986/9789610502838.

Fikfak, Jurij, Jože Prinčič, Jeffrey D. Turk, and Tatiana Bajuk Senčar. "To Be a Director in Time of Socialism: Between Ideas and Practice." In Fikfak, Prinčič, and Turk, *Biti direktor v času socializma,* 267–77.

"Financial Reports." Uljanik (website). Archived 6 June 2021 at the Wayback Machine. https://web.archive.org/web/20210616174903/https:/uljanik.hr/en/other-information/financial-reports.

Fonović Cvijanović, Teodora, and Vanessa Vitković Marčeta. "The Language of the Socialist Man: The Case of Istrian Periodicals." *Südost-Forschungen*, no. 76 (2017): 43–63. https://doi.org/10.1515/sofo-2017-760106.

Foucault, Michel. "Of Other Spaces." *Diacritics* 16, no. 1 (Spring 1986): 22–7. https://doi.org/10.2307/464648.

Gajdzik, Bożena. "Steel Industry in Poland: Trends in Production, Employment and Productivity in the Period from 2004 to 2019." *Metalurgija* 60, nos. 1–2 (2021): 165–8.

Gardawski, Juliusz, ed. *Polacy pracujący a kryzys fordyzmu*. Warsaw: Scholar, 2009.

Ghodsee, Kristen, and Mitchell Orenstein. *Taking Stock of Shock: Social Consequences of the 1989 Revolution*. Oxford: Oxford University Press, 2021. https://doi.org/10.1093/oso/9780197549230.001.0001.

Giddens, Anthony. *The Constitution of Society: Outline of the Theory of Structuration*. Cambridge: Polity Press, 1984.

Gligorov, Vladimir. "The Social and Economic Basis of Socialist Self-Management in Yugoslavia." In *The Functioning of the Yugoslav Economy*, edited by Radmila Stojanović, 3–22. Armonk, NY: M.E. Sharpe, 1982. https://doi.org/10.4324/9781315086552-1.

Goffman, Erving. *Asylums: Essays on the Social Situation of Mental Patients and Other Inmates*. New York: Anchor Books, 1961.

Gourdon, Karin. "An Analysis of Market-Distorting Factors in Shipbuilding: The Role of Government Interventions." OECD Science, Technology and Industry Policy Papers 67, 2019. https://doi.org/10.1787/23074957.

Graber Majchrzak, Sarah. *Arbeit – Produktion – Protest. Die Leninwerft in Gdańsk und die AG "Weser" in Bremen im Vergleich (1968–1983)*. Cologne: Böhlau, 2021. https://doi.org/10.7788/9783412519193.

– "The Gdańsk Shipyard: Production Regime and Workers' Conflicts in the 1970s and 1980s in the People's Republic of Poland." In Varela, Murphy, and van der Linden, *Shipbuilding and Ship Repair Workers*, 365–96. https://doi.org/10.1515/9789048530724-013.

Grala, Dariusz. *Reformy gospodarcze w PRL (1982–1989). Próba uratowania socjalizmu*. Warsaw: Trio, 2005.

Grdešić, Marko. "Tranzicija, sindikati i političke elite u Sloveniji i Hrvatskoj." *Politička misao* 43, no. 4 (2006): 121–41.

Gregurek, Miroslav. "Stupanj i učinci privatizacije u Hrvatskoj." *Ekonomski Pregled* 52, nos. 1–2 (2001): 115–88.

Grüner, Stefan, and Sabine Mecking, eds. *Wirtschaftsräume und Lebenschancen. Wahrnehmung und Steuerung von sozialökonomischem Wandel in Deutschland 1945–2000*. Berlin: De Gruyter Oldenbourg, 2017. https://doi.org/10.1515/9783110523010.

Grzymala-Busse, Anna, and Pauline Jones Luong. "Reconceptualizing the State: Lessons from Post-communism." *Politics & Society* 30, no. 4 (August 2002): 529–54. https://doi.org/10.1177/0090591702030004002.

Hann, Christopher M., ed. *Postsocialism: Ideals, Ideologies, and Practices in Eurasia*. London: Routledge, 2002. https://doi.org/10.4324/9780203428115.

–, ed. *Property Relations: Renewing the Anthropological Tradition*. Cambridge: Cambridge University Press, 1998.

– *Repatriating Polanyi: Market Society in the Visegrád States*. Budapest: Central European University Press, 2019.

Hardy, Jane. *Poland's New Capitalism*. London: Pluto Press, 2009.

Hellema, Duco. *The Global 1970s: Radicalism, Reform, and Crisis*. London: Routledge, 2018. https://doi.org/10.4324/9780429464133.

High, Steven C. *Industrial Sunset: The Making of North America's Rust Belt, 1969–1984*. Toronto: University of Toronto Press, 2003. https://doi.org/10.3138/9781442620902.

Hodges, Andrew. "Between Emic and Etic: 'Systematic' and 'Creative' Destruction during the Croatian Shipbuilding Crisis." *History in Flux* 2, no. 2 (2020): 92–110. https://doi.org/10.32728/flux.2020.2.5.

– "Psychic Landscapes, Worker Organizing and Blame: Uljanik and the 2018 Croatian Shipbuilding Crisis." *Südosteuropa* 67, no. 1 (2019): 50–74. https://doi.org/10.1515/soeu-2019-0003.

Hölscher, Jens, Nicole Nulsch, and Johannes Stephan. "Ten Years after Accession: State Aid in Eastern Europe." *European State Aid Law Quarterly* 13, no. 2 (2014): 305–16.

Horvat, Branko. *The Yugoslav Economic System: The First Labor-Managed Economy in the Making*. White Plains, NY: International Arts and Sciences Press, 1976.

Hossain, K.A., N.M.G. Zakaria, and M.A.R. Sarkar. "SWOT Analysis of China Shipbuilding Industry by Third Eyes." *Procedia Engineering*, no. 194 (2017): 241–6. https://doi.org/10.1016/j.proeng.2017.08.141.

Hrobat Virloget, Katja, Catherine Gousseff, and Gustavo Corni, eds. *At Home but Foreigners: Population Transfers in 20th Century Istria*. Koper: University of Primorska, 2015.

Jarausch, Konrad, ed. *Das Ende der Zuversicht? Die siebziger Jahre als Geschichte*. Göttingen: Vandenhoeck & Ruprecht, 2008. https://doi.org/10.13109/9783666361531.

Javeline, Debra. *Protest and the Politics of Blame: The Russian Response to Unpaid Wages*. Ann Arbor: University of Michigan Press, 2003. https://doi.org/10.3998/mpub.17850.

Jessop, Bop. "Spatial Fixes, Temporal Fixes and Spatio-temporal Fixes." In *David Harvey: A Critical Reader*, edited by Noel Castree and Derek Gregory, 142–66. Oxford: Blackwell, 2006. https://doi.org/10.1002/9780470773581.ch8.

Kalb, Don. "Conversations with a Polish Populist: Tracing Hidden Histories of Globalization, Class, and Dispossession in Postsocialism (and Beyond)." *American Ethnologist* 36, no. 2 (May 2009): 207–23. https://doi.org/10.1111/j.1548-1425.2009.01131.x.
- "Introduction. Headlines of Nation, Subtexts of Class: Working-Class Populism and the Return of the Repressed in Neoliberal Europe." In *Headlines of Nation, Subtexts of Class: Working-Class Populism and the Return of the Repressed in Neoliberal Europe*, edited by Don Kalb and Gábor Halmai, 1–36. New York: Berghahn Books, 2011. https://doi.org/10.1515/9780857452047-002.

Kalouptsidi, Myrto. "Detection and Impact of Industrial Subsidies: The Case of Chinese Shipbuilding." *Review of Economic Studies* 85, no. 2 (April 2018): 1111–58. https://doi.org/10.1093/restud/rdx050.

Kania, Stanisław. "Przemówienie tow. Stanisława Kani a zakończenia obrad III Plenum." *Nowe drogi* 81, no. 10 (1981): 28–33.

Karpiński, Andrzej, Stanisław Paradysz, Paweł Soroka, and Wiesław Żółtkowski. *Jak powstawały i jak upadały zakłady przemysłowe w Polsce*. Warsaw: Muza, 2013.

Keat, Preston. "Fallen Heroes: Explaining the Failure of the Gdansk Shipyard, and the Successful Early Reform Strategies in Szczecin and Gdynia." *Communist and Post-communist Studies* 36, no. 2 (June 2003): 209–30. https://doi.org/10.1016/S0967-067X(03)00026-6.

Kersan-Škabić, Ines. "Brodogradnja u Europskoj Uniji i Hrvatskoj – realnost i izazovi." *Ekonomska misao i praksa* 18, no. 2 (2009): 373–96.

Kideckel, David. *Getting By in Postsocialist Romania: Labor, the Body, and Working-Class Culture*. Bloomington: University of Indiana Press, 2008.

Kofti, Dimitra. "'Communists' on the Shop Floor." *Focaal*, no. 74 (March 2016): 69–82. https://doi.org/10.3167/fcl.2016.740106.
- "Moral Economy of Flexible Production: Fabricating Precarity between the Conveyor Belt and the Household." *Anthropological Theory* 16, no. 4 (December 2016): 433–53. https://doi.org/10.1177/1463499616679538.

Kojanic, Ognjen. "Nostalgia as a Practice of the Self in Post-Socialist Serbia." *Canadian Slavonic Papers* 57, nos. 3–4 (2015): 195–212. https://doi.org/10.1080/00085006.2015.1090760.

Kokanović, Marina. "The Cost of Nationalism: Croatian Labor, 1990–1999." In Crowley and Ost, *Workers after Workers' States*, 141–57.
- "Croatian Labour Realities: 1990–1999." *South East Europe Review* 2, no. 3 (1999): 185–208.

Kornai, János. *Economics of Shortage*. Amsterdam: North-Holland, 1980.
- "Resource-Constrained versus Demand-Constrained Systems." *Econometrica* 47, no. 4 (July 1979): 801–19. https://doi.org/10.2307/1914132.
- *The Socialist System: The Political Economy of Communism*. Princeton, NJ: Princeton University Press, 1992.

- "The Soft Budget Constraint." *Kyklos* 39, no. 1 (February 1986): 3–30. https://doi.org/10.1111/j.1467-6435.1986.tb01252.x.
Kotkin, Stephen. *Uncivil Society: 1989 and the Implosion of the Communist Establishment*. New York: Modern Library, 2009.
Kozarzewski, Piotr. *Polityka właścicielska państwa w okresie transformacji systemowej. Próba syntezy*. Lublin: Wydawnictwo Uniwersytetu Marii Curie-Skłodowskiej, 2019.
Krastev, Ivan, and Stephen Holmes. *The Light That Failed: A Reckoning*. London: Allen Lane, 2019.
Król, Marcin. *Byliśmy głupi*. Kraków: Znak, 2015.
Kumar Lahiri, Ashok. "Yugoslav Inflation and Money (May 1991)." IMF Working Paper 91/50, 1991. https://ssrn.com/abstract=884821.
Lammers, Konrad. "Subventionen für die Schiffbauindustrie." Working Paper 211, Institut für Weltwirtschaft, Universität Kiel, 1984.
Lampe, John. *Yugoslavia as History: Twice There Was a Country*. Cambridge: Cambridge University Press, 1996.
Lane, David. *The Socialist Industrial State: Towards a Political Sociology of State Socialism*. London: Allen & Unwin, 1976.
Leszczyński, Adam. *Ludowa historia Polski*. Warsaw: W.A.B., 2020.
Leyk, Aleksandra, and Joanna Wawrzyniak. *Cięcia. Mówiona historia transformacji*. Warsaw: Krytyka Polityczna, 2020.
Lipanović, B., and K. Kašper. "Jučer, danas, sutra – naše brodogradnje." *Brodogradnja* 34, no. 1 (1986): 5–10.
Maddison, Angus. *The World Economy*. 2 vols. Paris: OECD, 2006. https://doi.org/10.1787/9789264022621-en.
Marković, Predrag. "Der Sozialismus und seine sieben S-Werte der Nostalgie." In *Zwischen Amnesie und Nostalgie. Die Erinnerung an den Kommunismus in Südosteuropa*, edited by Ulf Brunnbauer and Stefan Troebst, 151–64. Cologne: Böhlau, 2007.
Markulinčić, Hrvoje, and Armando Debeljuh, eds. *Uljanik 1856–2006*. Pula: Arsenal design, 2006.
Marody, Mirosława. *Jednostka po nowoczesności. Perspektywa socjologiczna*. Warsaw: Scholar, 2014.
Martinčević, Juraj. "Samoupravna normativna djelatnost organizacija udruženog rada u procesu primjene zakona o udruženom radu." *Journal of Information and Organizational Sciences*, no. 1 (1977): 211–28.
"Materijalni položaj jugoslavenske brodogradnje: Društvenom podrškom premostiti teškoće." *Brodogradnja* 28, no. 1 (1980): 9–13.
Matošević, Andrea. "Tehnička događajnica i radnička intima. Brodogradilište Uljanik u dokumentarnim filmovima *Kolos s Jadrana*, *Berge Istra* i *Godine hrđe*." *Etnološka tribina* 48, no. 41 (2018): 194–212. https://doi.org/10.15378/1848-9540.2018.41.07.

Mazurek, Małgorzata. *Społeczeństwo kolejki. O doświadczeniach niedoboru 1945–1989*. Warsaw: Trio, 2010.
Mihaljević, Domagoj. "The Political Framework of Industrial Meltdown in Croatia, 1990–2013." *METU Studies in Development* 41, no. 3 (December 2014): 349–70.
Milat, Božana, and Vesna Galić. "Croatian Shipbuilding Industry: Ships Delivered & on Order." *Brodogradnja* 45, no. 1 (1997): 15–25.
Ministerstwo Pracy i Polityki Socjalnej. *Pakt o przedsiębiorstwie państwowym w trakcie przekształcania*. Łódź: Fundacja Edukacyjna Przedsiębiorczości, 1992.
Mitchell, Katharyne. "Work Authority in Industry: The Happy Demise of the Ideal Type." *Comparative Studies in Society and History* 34, no. 4 (October 1992): 679–94. https://doi.org/10.1017/S0010417500018041.
Morris, Jeremy. *Everyday Post-Socialism. Working-Class Communities in the Russian Margins*. London: Palgrave, 2016. https://doi.org/10.1057/978-1-349-95089-8.
Muehlebach, Andrea. "The Body of Solidarity: Heritage, Memory, and Materiality in Post-industrial Italy." *Comparative Studies in Society and History* 59, no. 1 (January 2017): 96–126. https://doi.org/10.1017/S0010417516000542.
Murphy, Hugh. "China, Philippines, Singapore, Taiwan, and Vietnam." In Varela, Murphy, and van der Linden, *Shipbuilding and Ship Repair Workers*, 637–56. https://doi.org/10.1515/9789048530724-025.
– "Labour in the British Shipbuilding and Ship Repair Industries in the Twentieth Century." In Varela, Murphy, and van der Linden, *Shipbuilding and Ship Repair Workers*, 47–116. https://doi.org/10.1515/9789048530724-002.
Murphy, Hugh, and Stig Tenold. "Appendix 1: The Effects of the Oil Price Shocks on Shipbuilding in the 1970s." In Varela, Murphy, and van der Linden, *Shipbuilding and Ship Repair Workers*, 665–81. https://doi.org/10.1515/9789048530724-027.
Musić, Goran. *Making and Breaking the Yugoslav Working Class: The Story of Two Self-Managed Factories*. Budapest: Central European University Press, 2021. https://doi.org/10.7829/j.ctv1bvndfr.
– "Provincial, Proletarian, and Multinational: The Antibureaucratic Revolution in Late 1980s Priboj, Serbia." *Nationalities Papers* 47, no. 4 (July 2019): 581–96. https://doi.org/10.1017/nps.2018.29.
Nam, Hwasook. *Building Ships, Building a Nation: Korea's Democratic Unionism under Park Chung Hee*. Seattle: University of Washington Press, 2009.
Negt, Oskar, and Alexander Kluge. *Geschichte und Eigensinn*. 3 vols. Frankfurt am Main: Zweitausendeins, 1981.
Niethammer, Lutz. *Kollektive Identität. Heimliche Quellen einer unheimlichen Konjunktur*. Reinbek: Rowohlt, 2000.
Niethammer, Lutz, Alexander von Plato, and Dorothee Wierling. *Die volkseigene Erfahrung. Eine Archäologie des Lebens in der Industrieprovinz der DDR. 30 biographische Eröffnungen*. Reinbek: Rowohlt, 1991.

Nowak, Stefan. "System wartości społeczeństwa polskiego." *Studia socjologiczne* 4, no. 75 (1979): 155–73.

OECD Guidelines on Corporate Governance of State-Owned Enterprises. Paris: OECD, 2005. https://www.oecd.org/daf/ca/oecd-guidelines-corporate-governance-soes-2005.htm.

OECD Working Party on Shipbuilding. *Imbalances in the Shipbuilding Industry and Assessment of Policy Responses*. Paris: OECD, 2017. https://web-archive.oecd.org/2017-08-04/445879-Imbalances_Shipbuilding_Industry.pdf.

Offe, Claus. "Das Dilemma der Gleichzeitigkeit. Demokratisierung und Marktwirtschaft in Osteuropa." *Merkur* 45, no. 505 (1991): 279–92.

Orenstein, Mitchell. *Out of the Red: Building Capitalism and Democracy in Postcommunist Europe*. Ann Arbor: University of Michigan Press, 2001.

Ost, David. *The Defeat of Solidarity: Anger and Politics in Postcommunist Europe*. Ithaca, NY: Cornell University Press, 2005. https://doi.org/10.7591/9781501729270.

– "Workers and the Radical Right in Poland." *International Labor and Working Class History*, no. 93 (Spring 2018): 113–24. https://doi.org/10.1017/S0147547917000345.

Ostojić, Siniša. "Novi vjetrovi u jedrima poljske brodogradnje." *Brodogradnja* 43, no. 2 (1995): 151–2.

Otto, Wiesław. "Gra o wpływy. Gry i zabawy polskich przedsiębiorstw." *Przegląd Organizacji*, no. 10 (1986): 17–20.

Palairet, Michael. "Croatian Shipbuilding in Crisis, 1979–1995." In *Enterprise in Transition: Proceedings. Fourth International Conference on Enterprise in Transition, Split-Hvar, May 24–26, 2001*, edited by Srećko Goić, 758–818. Split, Croatia: University of Split Faculty of Economics, 2001.

Patel, Kiran, and Hans Christian Röhl. *Transformation durch Recht: Geschichte und Jurisprudenz europäischer Integration 1985–1992*. Tübingen: Mohr Siebeck, 2020. https://doi.org/10.1628/978-3-16-159021-4.

Pawlak, Mikołaj. "From Sociological Vacuum to Horror Vacui: How Stefan Nowak's Thesis Is Used in Analyses of Polish Society." *Polish Sociological Review*, no. 189 (2015): 5–27.

Peračković, Krešimir. "Promjene u strukturi zanimanja u Hrvatskoj od 1971. do 2001. Od ratara do konobara." *Sociologija i prostor* 45, nos. 3–4 (2007): 377–98.

Perić Hadžić, Ana, and Tea Karačić. "Restrukturiranje hrvatske brodogradnje u kontekstu pristupanja Europskoj Uniji." *Pomorski zbornik*, nos. 47–8 (2013): 121–32.

Petrović, Tanja. "'When We Were Europe': Socialist Workers in Serbia and Their Nostalgic Narratives. The Case of the Cable Factory Workers in Jagodina." In *Remembering Communism: Private and Public Recollections of Lived Experience in Southeast Europe*, edited by Maria Todorova, Augusta Dimou, and Stefan Troebst, 127–64. Budapest: Central European University Press, 2014.

Petrungaro, Stefano. "Ethics of Work and Discipline in Transition: Uljanik in Late and Post-Socialism." *Review of Croatian History*, no. 15 (2019): 191–213. https://doi.org/10.22586/review.v15i1.9803.

"PiS domaga się pomocy od Unii. Chce pieniędzy na stocznie." PortalMorski. pl, 17 October 2014. https://www.portalmorski.pl/stocznie-statki/27199 -pis-domaga-sie-pomocy-od-unii-chce-pieniedzy-na-stocznie.

Pittaway, Mark. *Eastern Europe, 1939–2000: Brief Histories*. London: Bloomsbury Academic, 2010.

Polanyi, Karl. *The Great Transformation. Politische und ökonomische Ursprünge von Gesellschaften und Wirtschaftssystemen*. Translated by Heinrich Jelinek. Frankfurt am Main: Suhrkamp, 1973.

– "Sozialistische Rechnungslegung." *Archiv für Sozialwissenschaft und Sozialpolitik* 49, no. 2 (1922): 377–420.

Portelli, Alessandro. *The Death of Luigi Trastulli and Other Stories: Form and Meaning in Oral History*. Albany: State University of New York Press, 1991.

– "What Makes Oral History Different." In *Oral History, Oral Culture, and Italian Americans*, edited by Luisa Del Giudice, 21–30. New York: Palgrave Macmillan, 2009. https://doi.org/10.1057/9780230101395_2.

Prinčič, Jože. "Direktorska funkcija v jugoslovanskem socialističnem gospodarskem sistemu." In Fikfak, Prinčič, and Turk, *Biti direktor v času socializma*, 57–103.

Przeworski, Adam. *Democracy and the Market: Political and Economic Reforms in Eastern Europe and Latin America*. New York: Cambridge University Press, 1991. https://doi.org/10.1017/CBO9781139172493.

Pula, Besnik. *Globalization under and after Socialism: The Evolution of Transnational Capital in Central and Eastern Europe*. Stanford: Stanford University Press, 2018. https://doi.org/10.11126/stanford/9781503605138 .001.0001.

Puljar D'Alessio, Sanja. *Mi gradimo brod, a brod gradi nas. Etnografija organizacije brodogradilišta 3. maj*. Zagreb: Institut za etnologiju i folkloristiku, 2018.

Raphael, Lutz. *Jenseits von Kohle und Stahl. Eine Gesellschaftsgeschichte Westeuropas nach dem Boom*. Berlin: Suhrkamp, 2019.

"Referendum o integraciji vodećih jadranskih brodogradilišta." *Brodogradnja* 18, no. 5 (1967): 242.

Reljanović, Mario. "The Normative Position of Trade Unions and the Ideals of Self-Management." In *We Have Built Cities for You: On the Contradictions of Yugoslav Socialism*, edited by Vida Knežević and Marko Miletić, 61–80. Belgrade: Center CZKD, 2018.

Ryšlavy, Branko. "Hrvatska brodogradnja danas. Razgovor s Karlom Radolovićem predsjednikom Uljanika." *Brodogradnja* 40, nos. 3–4 (1992): 103–9.

Sabel, Charles F., and David Stark. "Planning, Politics, and Shop-Floor Power: Hidden Forms of Bargaining in Soviet-Imposed State-Socialist Societies."

Politics & Society 11, no. 4 (December 1982): 439–75. https://doi.org
/10.1177/003232928201100403.
Sachs, Jeffrey, and David Lipton. "Poland's Economic Reform." *Foreign Affairs*
69, no. 3 (Summer 1990): 47–66. https://doi.org/10.2307/20044400.
Sajatović, Mijodrag. "Prenosimo." *Brodogradnja* 34, no. 3 (1986): 178–80.
"Sanacija i restrukturiranje hrvatskh brodogradilišta." *Brodogradnja* 47, no. 2
(1999): 111–15.
Santos, João Pedro. "From Iron to the Industrial Cloud: Memory and (De)
industrialization at the Lisnave and Setenave Shipyards." *Narodna umjetnost*
57, no. 1 (2020): 93–107. https://doi.org/10.15176/vol57no105.
Schierup, Carl-Ulrik. *Migration, Socialism and the International Division of
Labour: The Yugoslavian Experience*. Aldershot: Averbury, 1990.
Schönfelder, Bruno. "Schmerzlose Stabilisierung? Erfolge und Risiken der
kroatischen Stabilisierungspolitik." *Südosteuropa* 45, no. 2 (1996): 120–37.
Schuhmann, Annette. *Kulturarbeit im sozialistischen Betrieb. Gewerkschaftliche
Erziehungspraxis in der SBZ/DDR 1946 bis 1970*. Cologne: Böhlau, 2006.
Schult, Ulrike. *Zwischen Stechuhr und Selbstverwaltung. Eine Mikrogeschichte sozialer
Konflikte in der jugoslawischen Fahrzeugindustrie 1965–1985*. Münster: LIT, 2017.
Schulz, Ulrike, and Thomas Welskopp. "Wieviel kapitalistisches Unternehmen
steckte in den Betrieben des real existierenden Sozialismus? Konzeptionelle
Überlegungen und ein Fallbeispiel." *Jahrbuch für Wirtschaftsgeschichte* 58,
no. 2 (2017): 331–66. https://doi.org/10.1515/jbwg-2017-0013.
Schumann, Michael, Edgar Einemann, Christia Siebel-Rebell, and Klaus P.
Wittemann. *Rationalisierung, Krise, Arbeiter. Eine empirische Untersuchung
der Industrialisierung auf der Werft*. Frankfurt am Main: Europäische
Verlagsanstalt, 1982.
Schumpeter, Joseph A. *Capitalism, Socialism and Democracy*. New York:
Routledge, 2010. https://doi.org/10.4324/9780203857090.
Schuster, Leslie. *A Workforce Divided: Community, Labor, and the State in Saint-
Nazaire's Shipbuilding Industry, 1890–1910*. Santa Barbara, CA: Greenword, 2002.
Scribner, Charity. *Requiem for Communism*. Cambridge, MA: MIT Press, 2005.
SEA Europe. "Open Letter." 29 April 2021. https://www.seaeurope.eu
/images/SEA_Europe_Open_letter_29_April_2021_Final_copy.pdf.
Šesnić, Željko. "Posljednjih 60 godina Brodogradilišta Uljanik." In *Stotinu
i pedeset godina brodogradnje u Puli. Zbornik radova s međunarodnog skupa
prigodom 150. obljetnice osnutka C.kr. pomorskog arsenala (Pula, 8. prosinca
2006.)*, edited by Bruno Dobrić, 239–66. Pula, Croatia: Društvo za
proučavanje prošlosti C. i. kr. mornarice Viribus unitis, 2010.
Shin, Wonchul. "The Evolution of Labour Relations in the South Korean
Shipbuilding Industry: A Case Study of Hanjin Heavy Industries, 1950–2014."
In Varela, Murphy, and van der Linden, *Shipbuilding and Ship Repair Workers*,
615–35. https://doi.org/10.1515/9789048530724-024.

Shleifer, Andrei, and Daniel Treisman. "Normal Countries: The East 25 Years after Communism." *Foreign Affairs* 93, no. 6 (November/December 2014): 92–103. http://www.foreignaffairs.com/articles/142200/andrei-shleifer-and-daniel-treisman/normal-countries.

Siegrist, Hannes. "Perspektiven der vergleichenden Geschichtswissenschaft. Gesellschaft, Kultur, Raum." In *Vergleich und Transfer: Komparatistik in den Sozial-, Geschichts- und Kulturwissenschaften*, edited by Harmut Kaelble and Jürgen Schriewer, 307–28. Frankfurt am Main: Campus, 2003.

Sladovljev, Želimir. "Restrukturiranje hrvatske brodogradnje." *Brodogradnja* 42, no. 4 (1994): 289–91.

Slaven, Anthony. *British Shipbuilding, 1500–2010: A History*. Lancaster: Crucible, 2013.

Slobodian, Quinn. *Globalists: The End of Empire and the Birth of Neoliberalism*. Cambridge, MA: Harvard University Press, 2018. https://doi.org/10.4159/9780674919808.

Slowikowska, Karolina, and Patrick Graham. "Windmills Breathe Life into Failing Gdansk Shipyards." Reuters, 6 July 2012. https://www.reuters.com/article/us-poland-shipyard-idUSBRE8650CH20120706.

Song, Yann-huei Billy. "Shipping and Shipbuilding Policies in PR China." *Marine Policy* 14, no. 1 (January 1990): 53–70. https://doi.org/10.1016/0308-597X(90)90037-R.

Spaskovska, Ljubica, and Anna Calori. "A Nonaligned Business World: The Global Socialist Enterprise between Self-Management and Transnational Capitalism." *Nationalities Papers* 49, no. 3 (May 2021): 413–27. https://doi.org/10.1017/nps.2020.27.

Stanić, Igor. "'Jedan od najtežih dana u Uljaniku!' Štrajk u brodogradilištu Uljanik 1967. godine." *Problemi sjevernog Jadrana*, no. 15 (2016): 73–95. https://doi.org/10.21857/9xn31cvkzy.

Staniszkis, Jadwiga. *Poland's Self-Limiting Revolution*. Edited by Jan T. Gross. Princeton, NJ: Princeton University Press, 1984.

Stark, David. "Recombinant Property in East European Capitalism." *American Journal of Sociology* 101, no. 4 (January 1996): 993–1027. https://doi.org/10.1086/230786.

– *The Sense of Dissonance: Accounts of Worth in Economic Life*. Princeton, NJ: Princeton University Press, 2009. https://doi.org/10.1515/9781400831005.

Steidel, Christian, Laurent Daniel, and Cenk Yildiran. *Shipbuilding Market Developments Q2 2018*. Paris: OECD, 2018. https://web-archive.oecd.org/2018-09-03/492776-shipbuilding-market-developments-Q2-2018.pdf.

Stojčić, Nebojša. "The Two Decades of Croatian Transition: A Retrospective Analysis." *South East European Journal of Economics and Business* 7, no. 2 (November 2012): 63–76. https://doi.org/10.2478/v10033-012-0015-5.

Stopford, Martin. "Current and Past Policies for Expanding Maintaining or Reducing Shipbuilding Capacity." OECD Working Party 6, 9 November 2015. https://www.oecd.org/sti/ind/Item%203.3%20Stopford _ShipbuildingCapacity.pdf.
- *Maritime Economics*. 3rd ed. London: Routledge, 2009. https://doi.org/10.4324/9780203891742.

Stott, Paul. "Surviving EU Accession: The Seven Habits of Highly Effective Shipbuilders." In *Proceedings of the 18th Symposium on Theory and Practice of Shipbuilding (SORTA)*. Zagreb: University of Zagreb, Faculty of Mechanical Engineering and Naval Architecture, 2008. https://eprints.ncl.ac.uk/file_store/production/156380/EF63558D-A8FE-44B9-B45C-0804BDD11709.pdf.

Stråth, Bo. *The Politics of De-industrialisation: The Contraction of the West European Shipbuilding Industry*. London: Croom Helm, 1987.

Sučec, Nikola. "Kto je Gianni Rossanda, direktor Uljanika pod čijom je rukom pokleknuo istarski div." tportal, 29 August 2018. https://www.tportal.hr/biznis/clanak/tko-je-gianni-rossanda-direktor-uljanika-pod-cijom-je-rukom-pokleknuo-istarski-div-foto-20180829.

Sundhaussen, Holm. *Jugoslawien und seine Nachfolgestaaten 1943–2011. Eine ungewöhnliche Geschichte des Gewöhnlichen*. Vienna: Böhlau, 2014. https://doi.org/10.7767/boehlau.9783205793625.

Szczepański, Marcin R. *Konflikt przemysłowy w Polskich Kolejach Państwowych w latach 1989–2005. Podłoże, przebieg, znaczenie polityczne*. Toruń: Adam Marszałek, 2014.

Sznajder Lee, Aleksandra. *Transnational Capitalism in East Central Europe's Heavy Industry: From Flagship Enterprises to Subsidiaries*. Ann Arbor: University of Michigan Press, 2016. https://doi.org/10.3998/mpub.7480364.

Tarkowska, Elżbieta, and Jacek Tarkowski. "'Amoralny familizm,' czyli o dezintegracji społecznej w Polsce lat osiemdziesiątych." *Kultura i Społeczeństwo* 60, no. 4 (2016): 7–28. https://doi.org/10.35757/KiS.2016.60.4.1.

Ther, Philipp. *Die neue Ordnung auf dem alten Kontinent. Eine Geschichte des neoliberalen Europa*. Berlin: Suhrkamp, 2016.

Ther, Philipp. *Europe since 1989: A History*. Translated by Charlotte Hughes-Kreutzmüller. Princeton, NJ: Princeton University Press, 2016.
- *How the West Lost the Peace: The Great Transformation since the Cold War*. Translated by Jessica Spengler. Cambridge: Polity, 2023.
- "1989 und die globale Hegemonie des Neoliberalismus." In *Der Zusammenbruch der alten Ordnung? Die Krise der Sozialen Marktwirtschaft und der neue Kapitalismus in Deutschland und Europa*, edited by Christoph Lorke and Rüdiger Schmidt, 53–86. Stuttgart: Steiner, 2019.

Tholen, Jochen. "Traditional and New Fields for Shipyards' Activities: Some Selected Ideas." Presentation at the workshop "Firms in Late and Post-socialism," University of Pula, 27 September 2017.
Thompson, Paul. *The Voice of the Past: Oral History*. Oxford: Oxford University Press, 1978.
Todorova, Maria, ed. *Remembering Communism: Genres of Representation*. New York: Social Science Research Council, 2010.
Tooze, Adam. *Crashed: How a Decade of Financial Crises Changed the World*. London: Allen Lane, 2018.
Trappmann, Vera. *Fallen Heroes in Global Capitalism: Workers and the Restructuring of the Polish Steel Industry*. Houndmills, UK: Palgrave Macmillan, 2013. https://doi.org/10.1057/9781137303653.
Tutić, Željka. *The EU State Aid Rules: The Case of Croatia*. Budapest: Central European University, Department of Public Policy, 2011.
Tymiński, Maciej. "Managers in the Command Economy: Case Studies from Poland, 1956–1970." *Business History* 64, no. 1 (2022): 156–82. https://doi.org/10.1080/00076791.2019.1687686.
Tyson, Laura D'Andrea. "Liquidity Crises in the Yugoslav Economy: An Alternative to Bankruptcy?" *Soviet Studies* 29, no. 2 (April 1977): 284–95. https://doi.org/10.1080/09668137708411123.
"Ugovor o financiranju tri gradnje." *Brodogradnja* 48, no. 2 (2000): 112.
Uljanik. *Informativni Priručnik o osnovnim zanimanjima u brodogradniji*. Pula: Uljanik, 1981.
"'Uljanik' prvi izvoznik-proizvođač." *Brodogradnja* 24, no. 6 (1973): 396.
Unkovski-Korica, Vladimir. *The Economic Struggle for Power in Tito's Yugoslavia: From World War II to Non-alignment*. London: I.B. Tauris, 2016. https://doi.org/10.5040/9781350988606.
van der Linden, Marcel, Hugh Murphy, and Raquel Varela. "Introduction." In Varela, Murphy, and van der Linden, *Shipbuilding and Ship Repair Workers*, 15–44. https://doi.org/10.1515/9789048530724-001.
Vanhuysse, Pieter. *Divide and Pacify: Strategic Social Policies and Political Protests in Post-communist Democracies*. Budapest: Central European University Press, 2006. https://doi.org/10.1515/9786155211447.
Varela, Raquel, Hugh Murphy, and Marcel van der Linden, eds. *Shipbuilding and Ship Repair Workers around the World: Case Studies, 1950–2010*. Amsterdam: Amsterdam University Press, 2017. https://doi.org/10.1515/9789048530724.
Verdery, Katherine. *What Was Socialism, and What Comes Next?* Princeton, NJ: Princeton University Press, 1996. https://doi.org/10.1515/9781400821990.
Vodopivec, Frano, dir. *Berge Istra*. Zagreb: Jadran Film, 1972. YouTube video, 18:36. https://www.youtube.com/watch?v=GFCDLilfRUU.

Vodopivec, Nina. "Past for the Present: The Social Memory of Textile Workers in Slovenia." In Todorova, *Remembering Communism*, 213–36.

von Hirschhausen, Christian, and Jürgen Bitzer, eds. *The Globalization of Industry and Innovation in Eastern Europe: From Post-Socialist Restructuring to International Competitiveness*. Cheltenham, UK: Edward Elgar, 2000. https://doi.org/10.4337/9781782542308.

von Plato, Alexander. "Oral History als Erfahrungswissenschaft. Zum Stand der 'mündlichen Geschichte' in Deutschland." *BIOS* 4, no. 1 (1991): 97–119.

Warlouzet, Laurent. "The Collapse of the French Shipyard of Dunkirk and EEC State-Aid Control (1977–86)." *Business History* 62, no. 5 (2017): 858–78. https://doi.org/10.1080/00076791.2017.1307341.

Wegenschimmel, Peter. *Zombiewerften oder Hungerkünstler? Staatlicher Schiffbau in Ostmitteleuropa nach 1970*. Berlin: De Gruyter Oldenbourg, 2021. https://doi.org/10.1515/9783110736007.

Wolf, Johanna. "Bremer Vulkan: A Case Study of the West German Shipbuilding Industry and Its Narratives in the Second Half of the Twentieth Century." In Varela, Murphy, and van der Linden, *Shipbuilding and Ship Repair Workers*, 117–42. https://doi.org/10.1515/9789048530724-003.

Woodward, Susan. *Balkan Tragedy: Chaos and Dissolution after the Cold War*. Washington, DC: Brookings Institution Press, 1995.

World Trade Organization. *International Trade Statistics 2001*. Geneva: WTO, 2001. https://www.wto.org/english/res_e/statis_e/its2001_e/stats2001_e.pdf.

Żukowski, Tomasz. "Fabryki-urzędy. Rozważania o ładzie społeczno-gospodarczym w polskich zakładach przemysłowych w latach realnego socjalizmu." In *Zmierzch socjalizmu państwowego. Szkice z socjologii ekonomicznej*, edited by Witold Morawski. Warsaw: PWN, 1994.

Index

Note: The letter *f* following a page number denotes a figure; the letter *m*, a map.

3. Maj shipyard (Rijeka), 45, 73, 85–6, 88, 118

agency (of workers), 37, 157–8, 172, 174, 198, 200–1
agriculture, 20, 28, 133
alienation, 17, 146, 171, 181, 183, 198–9
Al Mokattam (cargo vessel), 52
Altmaier, Peter, 213–14
austerity, 9, 14, 18–19, 50, 99, 214
autonomy: corporate, 60–1, 67, 95–7, 99, 101, 115; of workers, 127, 161, 171, 173–4, 180, 198
Austria, 95, 208; Habsburg, 6

Balcerowicz, Leszek, 14, 25
bank: bad, 23, 85, 100; central/national, 49, 67, 96, 98, 100, 115; commercial, 63–72 *passim*, 75–8, 86, 98–9, 143; European Investment Bank, 122; foreign, 65, 170, 196; Istarska banka, 64, 98; Riječka banka, 64, 68, 77, 98; State Agency for Bank Rehabilitation, 108; World Bank, 210; Yugoslav Bank for International Economic Cooperation (Jubmes), 64
bankruptcy, 6–7, 24, 59, 96, 99, 120, 184–92; assets, 27, 34, 188; Lenin shipyard, 26, 73; Neptun shipyard, 83
basic organization of associated labour (OOUR), 58, 68, 99, 102–3, 137
Berge Istra (cargo vessel), 55–6, 177
Bilić, Karlo, 43, 58, 65
Brajković, Anton, 81
Bremen, Bremerhaven, 5, 83, 160, 166
Bretton Woods system, 49, 59
Bulić, Bruno, 140

capitalism, 55, 94, 128, 188; from below, 8; creative destruction, 16, 27, 88–9, 201; crony, 19–20, 109, 146–7, 159; embedded, 13; laissez-faire, 12, 98, 118; neoliberal, 9; state, 35; wild, 147
cargo vessel, 52, 54–5, 185, 192
celebration, 39*f*, 81, 175–7, 182. *See also* ceremony
ceremony, 150, 157, 175–7. *See also* celebration

262 Index

China (People's Republic), 15, 43–5, 52, 54, 80, 192–7, 213–14
China State Shipbuilding Corporation (CSSC, Cosco), 43, 45, 54, 80
Chinese Communist Party (CCP), 43
Civic Platform (PO), 120
clientelism, 109, 153
committee: disciplinary, 173; planning, 105; trade union, 140; work, 95
communism, 50, 108, 154, 158, 164–5, 203–5
company apartments, 75, 95, 135m–136, 139–40, 143, 187–8
competitiveness, 62, 81, 177, 185–6, 204–6; China, 194; EU, 121, 210
"competitive signaling," 20–1, 94, 205, 207
complex organization of associated labour (SOUR), 58, 103
computers, 50, 68–9, 115, 195
constitution (Yugoslavia), 8, 57, 98, 102, 115, 146
consumer goods, 30, 48, 88, 133; consumerism, 188
core business, 25, 29, 127, 144, 188, 206
corporatism, 32, 106
Council for Mutual Economic Assistance (Comecon), 24, 52
Covid-19 pandemic, 35, 42, 53, 209, 211, 213–14
Crist shipyard, 27, 89–90, 148–9, 189f
Croatian Democratic Union (HDZ), 19, 107–9, 119, 169
Croatian Shipbuilding (association). See Jadranbrod
cruise ship, 26, 54, 80, 86, 89–90
culture, 48, 128–37, 141, 155, 189, 191; subculture, 129–30; work, 161, 172, 188
currency: convertible/foreign, 54, 60–7, 105, 114, 204; national, 60, 170

customers: domestic, 49, 52–3, 62–4, 80; foreign/hard, 24, 51–3, 62–4, 67, 69–71, 80, 86
Čuvalo, Milan, 79, 116
Czechoslovakia / Czech Republic, 17, 19, 21

debt: conversion into shares, 20, 72, 107; foreign, 9, 19, 65–7, 78, 114; relief, 67, 70, 72, 77, 81–7, 99–100
decline (narrative of), 14–16, 126, 166, 190, 200–1
deindustrialization, 14–16, 20, 27, 163, 170, 188
Democratic Left Alliance (SLD), 109–11
democratization, 11, 17, 30, 122, 202
directors. See management
dock, 27, 55–7, 56f, 73, 112, 157, 181, 185, 193; dockworkers, 124, 131, 140, 177, 180
documentary film, 55, 175, 177, 181

East Asia, 53–4, 79–80, 121, 194–5, 209
economic crisis. See recession
economic growth. See GDP
economic policy: post-socialist, 16–19, 76, 107, 120, 208; under socialism, 50–1, 60, 67, 97
economic reform: under socialism, 57, 66, 95–7, 105–6; in transformation, 17–20, 94, 100, 108–11, 168–9, 207–8
economy: command, 10–11, 48, 60, 96, 105, 111, 114; global, 23, 35, 41, 49–52, 204–5; market, 11, 17, 98, 111, 121, 168; of shortage, 25, 127, 133–4, 141
education, 27–9, 143–4, 152, 180, 182, 184
engineers, 31, 143, 150, 152; as managers, 26, 183; post-1990, 7, 27, 90
engines, 43, 51f, 58, 68, 160, 177

ethnicity, 18, 154, 207
European Union, 13, 37, 42, 49, 194, 209–15; accession, 6–8, 82–5, 100–1, 119–23, 205, 207–11; aid/subsidies, 19, 110, 188, 193, 213; Commission, 13, 82–6, 110, 120–1, 209, 213; competition law, 46–8, 82–4, 110–11, 120–2, 192–3, 213–14; Council, 120–1; Green Deal, 214
export, 49–52, 60–9, 84, 114, 204

ferry, 26, 34, 64, 67, 73, 89
financialization, 23, 121, 184
financing (of new ships), 9, 64, 82, 93, 122
Fincantieri shipyard (Italy), 88
Finland, 26, 73, 89
flag of convenience (FOC) countries, 51, 64
food, 25, 133, 139–42, 214. See also under industry
Fordism, 126–7, 158–9, 161, 198, 200
foreign direct investment, 12, 18–19, 45, 82, 85, 89
foreign exchange (rate), 60, 65–6, 81–2, 170
foreign trade, 43, 51, 58, 63–4, 112–14, 204
France, 52, 63, 89, 96, 215
free trade, 12, 90, 194, 214–15
freighter, 55, 73, 121. See also cargo vessel
Fukuyama, Francis, 11

Gdańsk (city), 10, 31, 67, 71, 100, 124, 180, 193
GDP, 16–19, 50, 65–6, 70–1, 167, 169
Gdynia (city), 3–5, 31–2, 184–6, 190–1, 208–9
gender. See men; women
Germany, 13, 80, 88–9, 148, 193, 213; East, 17; West, 28, 53, 93, 96, 167

Gierek, Edward, 112–13, 133
Glasgow, 160, 166
globalization, 40–2, 46–9, 94, 194–5
Great Britain, 15, 21, 52–4, 193, 197
Guofeng, Hua, 43, 45

habitability, 160, 170, 187–9, 200
health care, 71, 76–7, 108, 137, 141–3, 145
hierarchy, 105, 172, 175, 179, 183
history: end of, 9, 11, 202; of labour / working class, 11, 31, 163, 168
holidays, 7, 25, 34, 78, 141, 145
holism, 30, 126, 149–50, 155, 158–60, 184
Hungary, 18–19, 147

Ibn Battuta (dredger), 211–12
identification, 30–2, 132, 155, 170–1, 181, 184–5
illiquidity, 64, 70, 72, 75–7, 100, 142
India, 52, 63
industrial heritage, 73, 183, 188, 190, 193
industrial policy: Asia vs. EU, 54–5, 194, 213–15; post-socialist, 73, 107–8, 111, 122; under socialism, 99
industry association, 61, 64, 102–7, 195
industry: food, 20, 127–8; heavy, 28, 50, 71–2, 127, 161–6, 195–6; light, 127, 155; textile, 20, 121, 156, 166, 177
inflation, 10, 59, 66, 71, 76, 78; EU, 214; global, 49, 65; hyperinflation, 18, 24, 78
informal relations, 71, 114, 137, 172, 184, 187
insolvency, 23, 41, 45, 81, 96, 118; state, 99, 205
International Monetary Fund (IMF), 19, 97, 99, 103, 117, 210
investors: domestic, 21, 82–3, 87, 90–1, 119. See also foreign direct investment

264 Index

Istria, 27–8, 69, 71, 90, 117, 167
Istrian Democratic Assembly (IDS), 90, 117
Italy, 88, 127, 163, 172, 179

Jadranbrod, 59, 63–4, 102–3, 106–7
Jan de Nul, 192, 211–12
Japan, 15f, 22, 53–4, 77, 80, 121
Jaruzelski, Wojciech, 66

Kashubia, 28, 95
Keqiang, Li, 45
Končar, Danko, 90, 119
Korea: North, 193; South, 15f, 54, 65, 80, 121, 162
Kornai, János, 59, 96–7, 115, 195, 210
Kuroń, Jacek, 18

labour costs, 19, 22, 195–6
labour migration, 30, 88, 185, 208. *See also* staff turnover
Latin America, 11, 63
Law and Justice Party (PiS), 13, 21, 30, 110, 120, 208
Law on Associated Labour (1976), 57, 68, 102–3, 115
layoffs, 50, 78, 84, 160, 197, 200
League of Communists of Yugoslavia (SKJ), 18, 32, 98, 129, 145, 173
Lenin shipyard (Gdańsk), 13, 24–6, 61, 67, 71–3, 84, 112
liberalism, 11, 13, 57, 120, 168, 208
licences, 51, 68
lifeworld, 126–7, 133, 155–9 *passim*, 167–70 *passim*, 190, 201
liquidation, 25, 83, 101, 110–11, 122, 199
losers (of transformation), 16, 117

Maciejewski, Zbigniew, 115, 183–4
management, 68, 103, 114–16, 146; appointment of directors/ managers, 21, 60, 73, 79, 97, 183–4; relations to workers, 32, 153, 179, 182
Marković, Ante, 18, 107, 117, 167
marszały (freighters), 55, 65; *Marshall Budyonny*, 55
martial law, 25, 50, 66, 96–7, 115, 124
meaning: loss of, 30, 166, 181, 185, 199; production of, 25, 127, 161–4, 182, 190–1; world of, 12, 37, 162–3, 170, 201
mechanical engineering, 20, 27, 100, 115
memory, 125–6, 130–3, 153–8, 163–7, 170; collective, 34, 166, 186, 189; community of shared, 140, 158, 163–4; musealization, 45, 185–6; places of, 31, 140, 153
men, 16, 36, 127, 154, 161, 172
military, 16, 31, 35, 71, 91, 137
mining, 20, 50, 114, 161, 170
Ministry of Finance, 77, 79, 82, 100, 119
Ministry of State Treasury, 20; replacement of Ministry for Ownership Transformation, 109–10
modernity: industrial, 16, 140, 161, 167, 171; post-industrial, 191; socialist, 126, 158, 185–6, 203

nationalism, 19, 30, 116–17, 169, 207
nationality. *See* ethnicity
nationalization, 21, 81, 86, 107, 116, 197; of solidarity, 98; in the West, 53
neoliberalism: in the EU, 121, 210; hegemony of, 9, 37, 50, 93–4, 207–8; loss of credibility, 35, 209, 214–15; post-socialist, 18–19, 33, 120, 159, 182, 205–8
newspaper, 3, 37–8, 69–70, 169; company, 103–4, 129, 147, 156, 170–1, 180; local, 34, 59, 114, 177

Non-aligned Movement, 10, 51, 54, 69
Norway, 52, 55, 89, 148–9
nostalgia, 125, 147, 163, 166–7, 185–91 *passim*, 199–200

occupational safety, 71, 137, 159, 198
oil crisis, 23–4, 49, 52, 60, 111, 195–6
Organisation for Economic Co-operation and Development (OECD), 84, 87, 121
ownership: employee, 21, 73, 85–6, 158; foreign, 93; municipal, 136; social, 86; state, 60, 77, 99, 106–10, 194, 205–6. *See also* residential property

parliamentary election, 19, 30, 109–10, 208–9
paternalism, 13, 92, 95, 152, 164
path dependency, 34, 41, 94, 106, 109, 122
penalty (contractual), 23, 62, 67–8, 82, 86–7, 96
pension, 78, 137, 187; funds, 20, 77, 86, 108; pensioner, 125, 141, 189
periodization, 8–11, 28–9, 49, 201–11
photograph, 92, 131, 153, 164, 167, 171–2
Plenković, Andrej, 45, 119
Polanyi, Karl, 11–17, 110–11, 122, 128, 153
Polish United Workers' Party (PZPR), 31, 95, 113*f*, 132
populism, 201; right-wing, 16, 30, 110, 208–9
Portugal, 16, 162
post-communists, 21, 109, 111
privatization, 17–21, 84–7, 100–1, 110, 158, 197–9; agency/fund, 107–9; of apartments, 30, 140, 197
production cost, 26, 59, 64, 82, 87
production facilities, 27, 45, 50, 68–9, 77–8, 91

productivity: post-socialist, 22, 72–7, 82, 85, 159, 181; under socialism, 29, 63, 116
profitability, 26, 65, 73–5, 78, 85–6, 144; importance of, 46–8, 69–70, 117–19, 158–9, 162
propaganda, 129, 147, 150, 162, 177
property (residential), 135*m*, 140, 143, 170
protectionism, 23, 99, 110, 119
protest: post-socialist, 19, 76, 87–8, 111, 169; under socialism, 31–2, 57, 124–5, 128, 164–6
Pula (city), 6, 130, 134–7, 154, 190–1, 209

Radolović, Karlo, 68–9, 79, 81, 107, 109, 114–18
Rakowski, Mieczysław, 24, 92
rationalization, 30, 180
recession: in 1970s/80s, 9–10, 49–53, 65–6, 121; in 1990s, 17, 139, 167; of 2008/9, 35, 79, 85–6, 208
recognition, 150, 175–7, 191, 199; loss of, 184, 198
redundancy pay, 78, 84, 196–7
reform communists, 18–19, 66, 99, 116
re-industrialization, 15, 196
religion, 18, 155, 168
remembrance culture, 31–2, 157, 161, 166–7. *See also* memory
repair shipyard: Remontowa (Gdańsk), 192, 197; Viktor Lenac (Rijeka), 80
resorts. *See* holidays
restructuring: economy, 19, 210; shipyards, 77–9, 82–7, 99–101, 105–8, 117–18, 141–4
retirement, 78, 137, 146, 182, 200, 206
reward system, 39, 87, 147
Rijeka (city), 45, 80, 85

Rossanda, Gianni, 87, 118, 184
Rostock, 5, 83, 90, 213
Russia, 42, 80, 186–7, 214

Sachs, Jeffrey, 11
schools, 124, 137, 180; vocational/technical, 27–9, 144, 152, 160
Schumpeter, Joseph, 16, 27, 88
security: sense of, 16, 30, 166–7; social (insurance), 72, 99–100, 137
self-management: Poland, 95–9; Serbia, 17–18, 156, 169, 171; Yugoslavia, 21, 29, 57–63, 103, 109, 137, 154
service sector, 5, 34, 177, 188, 195, 199
shares, 20–1, 72–3, 77, 81, 85–6, 108; shareholder, 21, 26, 72–5, 86–7, 107–10, 117, 158
ship launch, 5, 149–50, 157, 177, 184
shipping company, 53–5, 63–4, 79, 89, 96, 121. *See also* Uljanik Plovidba
Shipyards' & Maritime Equipment Association of Europe (SEA Europe), 213–14
shock therapy, 19–20, 93, 117, 160, 196, 207
Slovenia, 19, 63, 145, 156, 158, 179
sociability, 33, 129, 166–7
social protection (need for), 13, 98, 100, 110, 119–20, 152–3
social advancement, 28, 132
social democracy, 11, 14, 81, 117, 162, 209
social policy, 108, 117, 122, 208
soft budget constraints, 59, 65–6, 83, 96, 115, 194, 210
solidarity, 98–101, 127–8, 139, 150–9, 166–7, 179
Solidarność (trade union): post-socialist, 8, 13, 18, 111, 168–9; under socialism, 10, 29, 66, 95–6, 105–6, 155

Soviet Union, 24, 51–2, 55, 59–65, 71, 163
Split (city), 85, 87
sport, 34, 95, 124, 132, 134–41, 189
staff turnover, 95, 136–7
standard of living, 14–19 *passim*, 50, 133, 137–8, 151, 161–7
state aid, 50–4, 62, 78–9, 82–5, 96–100, 117–22, 204
state control, 20, 42 60, 82, 99–110 *passim*
status: political, 101, 112; social, 31, 33, 142, 167–8, 200
steel, 50, 62, 147–50, 161–4, 194–5, 210
Stena Germanica (passenger ship), 67
Sting (Gordon Matthew Thomas Sumner), 161, 163, 190–1
strike: post-socialist, 76, 87–8, 169; under socialism, 31–2, 61, 103, 124, 165, 179–80
structural change: in the East, 49–50, 54–5, 100–1, 159, 188, 195–204 *passim*. *See also* Western Europe: structural change
subcontractors, 27, 62, 142, 146, 149, 181–2
Sudan, 69–70
supplier, 30, 60–2, 70–2, 72, 90, 105, 113
supply chain, 60, 89–90, 150, 160, 193, 213
Supreme Court of Audit, 67, 96
Sweden, 23, 53, 67, 96, 151
Szczecin (city), 31, 124
Szlanta, Janusz, 21, 26, 73–5, 78, 81

tanker, 52–9 *passim*, 56f, 69, 73, 112, 121, 175–7
technology: import from West, 61, 66, 114; outdated, 21–2, 62–3, 71, 146, 197, 206
Tito, Josip Broz, 10, 43, 51, 57, 66, 177

tourism, 5–8, 27, 34–5, 90–1, 130, 136, 140–1, 169, 184–91
trade union: post-socialist, 72, 76–9, 84–5, 111, 168–9, 208–10; under socialism, 124, 144–6; in West, 196
training. *See* education
transition, 11, 49, 168, 195–6, 205
Tuđman, Franjo, 19, 107, 116, 169

Uljanik Plovidba, 23, 69, 85
unemployment, 5, 18–19, 29–30, 71, 145, 170
United States, 14, 65, 188, 196–7, 208, 215

Veruda, Sale (Saša Milanović), 124, 129
violence, 31, 66, 124
Vrandečić, Ivo, 59, 64

wages, 19, 28–9, 72, 76, 161, 198; deductions, 137–40; minimum, 45, 88, 208; outstanding, 24, 76, 82, 87–8, 142, 169; wage gap/spread, 29, 33, 147, 182
Wallsend, 161, 163
war: Cold, 18; of Independence (Croatia), 19, 24–5, 71, 75–6, 116–18, 169; Russian against Ukraine, 42, 214; Second World, 18, 154
Warsaw, 18, 111, 132, 209
Warski shipyard (Szczecin), 13, 71–2, 81, 83–4, 110, 112
Washington Consensus, 18, 37

welder, 33, 144, 147, 164, 178
welfare: enterprise-based, 70, 127–9, 133–7, 141–3, 155–6, 164–6; state, 11, 13, 50, 78
Western Europe: shipbuilding, 22, 52–4, 76; structural change, 49–50, 78, 96, 196, 210
Willem van Rubroeck (dredger), 86, 192, 211
women, 28–9, 36, 78, 127–8, 140, 153–4, 158, 164, 166, 175, 178–9, 198
work discipline, 71, 147, 153, 159, 173–5, 181–2
work organization (RO), 58, 68, 99, 102–3, 171, 173
worker dynasty, 29, 34, 131, 161
worker: agency/temporary, 55, 181; blue-collar, 29, 155, 172, 208; skilled/specialist, 31, 33, 90, 171, 173; white-collar, 28–9, 31, 155, 172, 179
workers' council, 17, 32, 58, 68, 97, 115
workforce (reduction of), 27, 59, 77, 82, 84–5, 160
working class: decline of, 93, 163, 168–9, 208–9; fragmentation of, 31, 33, 154; idealization of, 33, 112, 150, 162, 164–5

youth, 30, 34, 76, 132, 171

Zagreb, 27, 88, 103, 143

GERMAN AND EUROPEAN STUDIES

General Editor: James Retallack

1 Emanuel Adler, Beverly Crawford, Federica Bicchi, and Rafaella Del Sarto, *The Convergence of Civilizations: Constructing a Mediterranean Region*
2 James Retallack, *The German Right, 1860–1920: Political Limits of the Authoritarian Imagination*
3 Silvija Jestrovic, *Theatre of Estrangement: Theory, Practice, Ideology*
4 Susan Gross Solomon, ed., *Doing Medicine Together: Germany and Russia between the Wars*
5 Laurence McFalls, ed., *Max Weber's 'Objectivity' Revisited*
6 Robin Ostow, ed., *(Re)Visualizing National History: Museums and National Identities in Europe in the New Millennium*
7 David Blackbourn and James Retallack, eds., *Localism, Landscape, and the Ambiguities of Place: German-Speaking Central Europe, 1860–1930*
8 John Zilcosky, ed., *Writing Travel: The Poetics and Politics of the Modern Journey*
9 Angelica Fenner, *Race under Reconstruction in German Cinema: Robert Stemmle's Toxi*
10 Martina Kessel and Patrick Merziger, eds., *The Politics of Humour: Laughter, Inclusion, and Exclusion in the Twentieth Century*
11 Jeffrey K. Wilson, *The German Forest: Nature, Identity, and the Contestation of a National Symbol, 1871–1914*
12 David G. John, *Bennewitz, Goethe,* Faust: *German and Intercultural Stagings*
13 Jennifer Ruth Hosek, *Sun, Sex, and Socialism: Cuba in the German Imaginary*
14 Steven M. Schroeder, *To Forget It All and Begin Again: Reconciliation in Occupied Germany, 1944–1954*
15 Kenneth S. Calhoon, *Affecting Grace: Theatre, Subject, and the Shakespearean Paradox in German Literature from Lessing to Kleist*
16 Martina Kolb, *Nietzsche, Freud, Benn, and the Azure Spell of Liguria*
17 Hoi-eun Kim, *Doctors of Empire: Medical and Cultural Encounters between Imperial Germany and Meiji Japan*
18 J. Laurence Hare, *Excavating Nations: Archeology, Museums, and the German-Danish Borderlands*
19 Jacques Kornberg, *The Pope's Dilemma: Pius XII Faces Atrocities and Genocide in the Second World War*
20 Patrick O'Neill, *Transforming Kafka: Translation Effects*
21 John K. Noyes, *Herder: Aesthetics against Imperialism*

22 James Retallack, *Germany's Second Reich: Portraits and Pathways*
23 Laurie Marhoefer, *Sex and the Weimar Republic: German Homosexual Emancipation and the Rise of the Nazis*
24 Bettina Brandt and Daniel L. Purdy, eds., *China in the German Enlightenment*
25 Michael Hau, *Performance Anxiety: Sport and Work in Germany from the Empire to Nazism*
26 Celia Applegate, *The Necessity of Music: Variations on a German Theme*
27 Richard J. Golsan and Sarah M. Misemer, eds., *The Trial That Never Ends: Hannah Arendt's* Eichmann in Jerusalem *in Retrospect*
28 Lynne Taylor, *In the Children's Best Interests: Unaccompanied Children in American-Occupied Germany, 1945–1952*
29 Jennifer A. Miller, *Turkish Guest Workers in Germany: Hidden Lives and Contested Borders, 1960s to 1980s*
30 Amy Carney, *Marriage and Fatherhood in the Nazi SS*
31 Michael E. O'Sullivan, *Disruptive Power: Catholic Women, Miracles, and Politics in Modern Germany, 1918–1965*
32 Gabriel N. Finder and Alexander V. Prusin, *Justice behind the Iron Curtain: Nazis on Trial in Communist Poland*
33 Parker Daly Everett, *Urban Transformations: From Liberalism to Corporatism in Greater Berlin, 1871–1933*
34 Melissa Kravetz, *Women Doctors in Weimar and Nazi Germany: Maternalism, Eugenics, and Professional Identity*
35 Javier Samper Vendrell, *The Seduction of Youth: Print Culture and Homosexual Rights in the Weimar Republic*
36 Sebastian Voigt, ed., *Since the Boom: Continuity and Change in the Western Industrialized World after 1970*
37 Olivia Landry, *Theatre of Anger: Radical Transnational Performance in Contemporary Berlin*
38 Jeremy Best, *Heavenly Fatherland: German Missionary Culture and Globalization in the Age of Empire*
39 Svenja Bethke, *Dance on the Razor's Edge: Crime and Punishment in the Nazi Ghettos*
40 Kenneth S. Calhoon, *The Long Century's Long Shadow: Weimar Cinema and the Romantic Modern*
41 Randall Hansen, Achim Saupe, Andreas Wirsching, and Daqing Yang, eds., *Authenticity and Victimhood after the Second World War: Narratives from Europe and East Asia*
42 Rebecca Wittmann, ed., *The Eichmann Trial Reconsidered*
43 Sebastian Huebel, *Fighter, Worker, and Family Man: German-Jewish Men and Their Gendered Experiences in Nazi Germany, 1933–1941*

44 Samuel Clowes Huneke, *States of Liberation: Gay Men between Dictatorship and Democracy in Cold War Germany*
45 Tuska Benes, *The Rebirth of Revelation: German Theology in an Age of Reason and History, 1750–1850*
46 Skye Doney, *The Persistence of the Sacred: German Catholic Pilgrimage, 1832–1937*
47 Matthew Unangst, *Colonial Geography: Race and Space in German East Africa, 1884–1905*
48 Deborah Barton, *Writing and Rewriting the Reich: Women Journalists in the Nazi and Post-war Press*
49 Martin Wagner, *A Stage for Debate: The Political Significance of Vienna's Burgtheater, 1814–1867*
50 Andrea Rottmann, *Queer Lives across the Wall: Desire and Danger in Divided Berlin, 1945–1970*
51 Jeffrey Schneider, *Uniform Fantasies: Soldiers, Sex, and Queer Emancipation in Imperial Germany*
52 Alexandria N. Ruble, *Entangled Emancipation: Women's Rights in Cold War Germany*
53 Johanna Schuster-Craig, *One Word Shapes a Nation: Integration Politics in Germany*
54 Kiran Klaus Patel, ed., *Tangled Transformations: Unifying Germany and Integrating Europe, 1985–1995*
55 Abraham Rubin, *Conversion and Catastrophe in German-Jewish Émigré Autobiography*
56 Ulf Brunnbauer, Philipp Ther, Piotr Filipkowski, Andrew Hodges, Stefano Petrungaro, and Peter Wegenschimmel, *In the Storms of Transformation: Two Shipyards between Socialism and the EU*

www.ingramcontent.com/pod-product-compliance
Lightning Source LLC
Chambersburg PA
CBHW020359080526
44584CB00014B/1092